Single-Molecule Optical Detection, Imaging and Spectroscopy

edited by
T. Basché, W. E. Moerner,
M. Orrit, U. P. Wild

VCH

© VCH Verlagsgesellschaft mbH, D-69451 Weinheim (Federal Republic of Germany), 1997

Distribution:
VCH, P.O. Box 10 11 61, D-69451 Weinheim (Federal Republic of Germany)
Switzerland: VCH, P.O. Box, CH-4020 Basel (Switzerland)
United Kingdom and Ireland: VCH (UK) Ltd., 8 Wellington Court, Cambridge CB1 1HZ
(England)
USA and Canada: VCH Publishers, Inc., 337 7th Avenue, New York, NY 10001 (USA)
Japan: VCH, Eikow Building, 10-9 Hongo 1-chome, Bunkyo-ku, Tokyo 113 (Japan)

ISBN 3-527-29316-7

Single-Molecule Optical Detection, Imaging and Spectroscopy

edited by
T. Basché,
W. E. Moerner,
M. Orrit,
U. P. Wild

Weinheim • New York •
Basel • Cambridge • Tokyo

T. Basché
Institut für Physikalische Chemie
Ludwig-Maximilians-Universität
Sophienstraße 11
D-80333 München

W. E. Moerner
Department of Chemistry and Biochemistry
University of California, San Diego
9500 Gilman Drive
La Jolla, CA 92093-0340
USA

M. Orrit
Université Bordeaux I
Centre de Physique Moléculaire
Optique et Hertzienne
351 Cours de Libération
F-33405 Talence

U. P. Wild
ETH Zürich
Laboratorium für Physikalische Chemie
Universitätsstraße 22
CH-8092 Zürich

Published by
VCH Verlagsgesellschaft mbH, Weinheim (Federal Republic of Germany)

Editorial Director: Dr. Michael Bär
Production Manager: Dipl.-Ing. (FH) Hans Jörg Maier

Library of Congress Card No. applied for.
A catalogue record for this book is available from the British Library.

Deutsche Bibliothek Cataloguing-in-Publication Data:
Single-Molecule Optical Detection, Imaging and Spectroscopy / ed. by T. Basché ... -
Weinheim ; New York ; Basel ; Cambridge ; Tokyo ; VCH, 1996
  ISBN 3-527-29316-7
NE: Basché, Thomas [Hrsg.]

Composition: Asco Trade Typesetting Limited, Hong Kong. Printing: betz druck gmbh, D-64291 Darmstadt. Bookbinding: J. Schäffer GmbH & Co. KG, D-67269 Grünstadt.
Printed in the Federal Republic of Germany.

Preface

Techniques for detection and controlled manipulation of single particles such as atoms, molecules, proteins or nanocrystals have been continuously emerging during the past ten years. Scanning tunneling microscopy and atomic force microscopy allowed for the first time the imaging of single atoms and molecules on surfaces in real space. The optical detection of single particles was first realized by imaging the fluorescence of single atomic ions stored in radiofrequency traps. The present book is dedicated to a survey of optical methodologies to detect and image single organic dye molecules (single fluorophores) in solids, on surfaces and in liquids.

Optical experiments at the single-molecule level hold promise for novel and unexpected achievements in different fields of science. In a single-molecule experiment the usual averaging over large populations is absent and inhomogeneous distributions of different origins that complicate measurements on large ensembles are eliminated. Various kinds of spectroscopy and microscopy and some clever combinations of both can be employed to detect single molecules in the condensed phase between liquid helium temperatures and room temperature. The range of techniques and experimental conditions available spans a very broad research arena, which includes quantum optics, the probing of dynamical interactions in solids and on surfaces on a nanoscopic scale, trace analysis and rare event screening in liquids, as well as studies of molecular processes in systems of biological interest.

The wide range of potential applications demonstrates that the topic of single-molecule detection and spectroscopy is quite interdisciplinary. As such, the book is intended for researchers, graduate students and advanced undergraduates in the field of chemical physics, solid-state physics, analytical chemistry, laser spectroscopy, photochemistry and photophysics of molecules, fluorescence spectroscopy, fluorescence microscopy and molecular biology. The book provides thorough introductions to the various methodologies, which makes it also useful for newcomers who wish to enter the field. Since optical experiments at the single-molecule level are a very rapidly expanding area of research a comprehensive list of references is available at the end of each section of the book.

The book is organized according to the different techniques that are used for single-molecule detection. Chapter 1 which is divided into six sections is devoted to single molecule studies in solids at low temperature. Section 1.1 introduces the field by reviewing the physical principles, methods and experimental techniques of high-resolution spectroscopy of single impurity molecules in solids. Section 1.2 treats the single-molecule excitation lineshape, dispersed fluorescence spectra and quantum optical experiments. Fluorescence microscopy, lifetime measurements, polarization effects and external perturbations by electric fields or hydrostatic pressure are discussed in section 1.3. The absorption lines of single molecules in solids at low temperatures often undergo frequency jumps (spectral shifts). This behaviour is described in section 1.4, and analyzed theoretically in section 1.5. The last section (1.6) of chapter 1 is devoted to magnetic resonance experiments on single molecular spins with or without an applied magnetic field and to the study of spin coherence.

In chapters 2 and 3 microscopic techniques suitable for single-molecule detection are described. Chapter 2 portrays near-field optical microscopy, a technique with sub-diffraction limited spatial resolution. A qualitative introduction into the basics of near-field optical microscopy is given as well as a synopsis of the various results that were achieved in single-molecule imaging and spectroscopy using this technique. The book concludes with a survey of the potential of single-molecule detection in analytical chemistry (chapter 3). Several microscopic techniques for single-molecule fluorescence detection in solution are outlined followed by applications of single-molecule detection in DNA sequencing and capillary electrophoresis.

The editors hope that the present volume will furnish the reader with a thorough understanding of the basic principles of the very rapidly expanding field of single-molecule detection and spectroscopy in the condensed phase. We would be particularly delighted if the book would also serve yet another purpose, namely to stimulate new research directions in this exciting area. We are grateful to S. Mais for technical assistance and to Dr. T. Mager and Dr. M. Baer from VCH publishers for their assistance and constructive help in preparing this book.

October 1996

T. Basché
W. Moerner
M. Orrit
U. P. Wild

List of Contents

List of Contributors

W. E. Moerner
Department of Chemistry
and Biochemistry
University of California, San Diego
9500 Gilman Drive
La Jolla, CA 92093-0340
USA

T. Basché
Ludwig-Maximilians-Universität
München
Institut für Physikalische Chemie
Sophienstraße 11
D-80333 München
Germany

S. Kummer
Ludwig-Maximilians-Universität
München
Institut für Physikalische Chemie
Sophienstraße 11
D-80333 München
Germany

C. Bräuchle
Ludwig-Maximilians-Universität
München
Institut für Physikalische Chemie
Sophienstraße 11
D-80333 München
Germany

M. Croci
ETH Zürich
Universitätsstraße 22
CH-8022 Zürich
Switzerland

H.-J. Müschenborn
ETH Zürich
Universitätsstraße 22
CH-8022 Zürich
Switzerland

U. P. Wild
ETH Zürich
Universitätsstraße 22
CH-8022 Zürich
Switzerland

R. Brown
Université Bordeaux I
Centre de Physique Moléculaire
Optique et Hertzienne
351 Cours de Libération
F-33405 Talence
France

M. Orrit
Université Bordeaux I
Centre de Physique Moléculaire
Optique et Hertzienne
351 Cours de Libération
F-33405 Talence
France

J.L. Skinner
University of Wisconsin
Department of Chemistry
Theoretical Chemistry Institute
Madison, WI 53706
USA

J. Wachtrup
Technische Universität
Chemnitz-Zwickau
Institut für Physik
D-09107 Chemnitz
Germany

C. von Borczyskowski
Technische Universität
Chemnitz-Zwickau
Institut für Physik
D-09107 Chemnitz
Germany

J. Köhler
Heinrich-Heine-Universität
Lehrstuhl für Festkörperspektroskopie
Universitätsstraße 1
D-40255 Düsseldorf
Germany

J. Schmidt
Rijksuniversiteit Leiden
Huygenslaboratorium
P. O. Box 9504
2300 RA Leiden
The Netherlands

J. Trautmann
SEQ Ltd.
201 Washington Road
Princeton, NJ 08540
USA

W. P. Ambrose
Los Alamos National Laboratory
Mail Stop M888
Los Alamos, NM 87545
USA

N. J. Dovichi
University of Alberta
Department of Chemistry
Edmonton, Alberta T6G 2G2
Canada

D. D. Chen
University of Alberta
Department of Chemistry
Edmonton, Alberta T6G 2G2
Canada

1 Low-Temperature Studies in Solids

1.1 Physical Principles and Methods of Single-Molecule Spectroscopy in Solids

W. E. Moerner

1.1.1 Introduction – why do single-molecule studies in solids?

Over the past few years, the power of optical spectroscopy with high-resolution laser sources has been extended into the fascinating domain of individual impurity molecules in solids. In this regime, the single molecule acts as an exquisitely sensitive probe of the details of the immediate local environment (which may be termed the "nanoenvironment") [1–8]. Using techniques described in this section and illustrated in the rest of Chapter 1 of this book, exactly one molecule hidden deep within a solid sample can now be probed at a time by tunable laser radiation, which represents detection and spectroscopy at the ultimate sensitivity level of 1.66×10^{-24} moles of material, or 1.66 yoctomole.[1]

Optical experiments in this new regime are generating much interest for a variety of reasons. Most importantly, single-molecule measurements completely remove the normal ensemble averaging that occurs when a large number of molecules are probed at the same time. Thus, the usual assumption that all molecules contributing to the ensemble average are identical can now be directly examined on a molecule-by-molecule basis. On the theoretical side, since no ensemble averaging need be done before computing an observable quantity, stronger tests of truly microscopic dynamical theories can be completed. Finally, since this is a previously unexplored regime, new physical and chemical behavior is likely to be observed, and many examples of this are presented in the chapters to come.

Single-molecule spectroscopy (SMS) in solids is related to, but distinct from, the fascinating and well-established field of spectroscopy of single electrons or ions confined in electromagnetic traps [10–12]. The vacuum environment and confining fields of an electromagnetic trap are quite different from the environment of a single molecule in a solid. The trap experiments must deal with micromotion in the confining trap potential and, to date, no single molecule has been cooled sufficiently to be bound by an electromagnetic trap. In SMS, however, the interactions with the lattice act to constrain the molecule, hindering or preventing molecular rotation. At the same time, the single molecule is continuously bathed in the phonon vibrations of the

---

[1] Since a single molecule is the smallest unit of a molecular substance, a more appropriate unit in this case would be the guacamole, which is the quantity of moles exactly equal to the inverse of avocado's number [9]. (With apologies to Amadeo Avogadro)

solid available at a given temperature, and can interact with the electric, magnetic and strain fields of the nanoenvironment.

Useful comparison may also be made to another important field, the direct probing of atoms or molecules on surfaces with scanning tunneling microscopy (STM) [13] or atomic force microscopy (AFM) [14, 15]. In STM and AFM of single molecules, a fairly strong bond must exist between the molecule and the underlying surface in order for the molecule to tolerate the perturbing forces from the tunnelling electrons or the tip. Of course, the spatial resolution of these methods is much higher as a result of the relatively short electron wavelength of a few tenths of a nanometer, and the tunneling or force tip must be placed correspondingly close to the molecule, and the molecule must be on the surface. On the other hand, SMS usually operates noninvasively in the optical far-field with a corresponding loss in spatial resolution to a value on the order of the optical wavelength (1 µm), but with no loss in spectral resolution. Moreover, single molecules can be studied below the surface in the body of the sample, and different single molecules can be selected by simply changing the optical wavelength used. In 1993, single-molecule imaging was achieved with near-field *optical* techniques with 100 nm resolution at room temperature [16, 17]. These near-field optical studies at room temperature and other near-field studies at liquid helium temperatures [18] will be described in chapter 2 of this book.

This section presents an overview of the physical principles and methods of high-resolution spectroscopy of single impurity centers in solids. The emphasis is on single, isolated molecular impurities, although many of the concepts described are also applicable to the possible future study of single ions as well [19]. The presentation will concentrate on the far-field regime, in which the spatial resolution is limited by diffraction effects to beam diameters on the order of 1 µm. Chapter 2 of this book treats the optical near-field, sub-wavelength realm, and Chapter 3 describes single-molecule detection in liquids. Section 1.1.2 describes the fundamental requirements for high-resolution SMS [20]. Section 1.1.3 describes the various experimental methods used to achieve SMS in solids, with selected examples. In the remaining sections of Chapter 1, specific results on external perturbations [21, 22], microscopic imaging [23], spectral shifting (also called spectral diffusion [3, 24], optical modifications of single molecules (which may eventually lead to a single-molecule optical storage) [25], correlation properties of the emitted photons [26, 27], quantum optical effects [28, 29], vibrational modes [30, 31], and single-spin magnetic resonance [32, 33] will be described in more detail. The reader may consult one of several recent reviews for more information [5–8].

The significant features of these SMS studies are: (i) new, unexpected physical effects have been observed in the single-molecule regime as a result of the nano-environmental sensitivity of the single-molecule lineshape, (ii) it is now possible to probe the members of the usual ensemble average *one at a time*, and therefore to directly measure the distribution rather than only measure its moments, (iii) as a result of the ability to follow spectral changes of a single molecule in real time, it is now possible in a single nanoenvironment to directly probe the connection between specific microscopic theories [34, 35] of local structure, dynamics, and host–guest interactions and the statistical mechanical averages that are measured in conventional experiments, (iv) the door to measurements on a single molecular spin has been opened for the first time, which should lead to unprecedented information on local magnetic interactions, and (v) quantum optical studies may now be performed in solids where the normal transit time or micromotion effects present in trap or beam experiments are absent.

1.1.2 Physical principles and optimal conditions

1.1.2.1 General considerations

One may ask: how is it possible to use optical radiation to isolate a single impurity molecule hidden deep inside a host matrix? To answer this question concisely, single-molecule optical spectroscopy is accomplished by selecting experimental conditions such that only one molecule is in resonance in the volume probed at a time. More precisely, a combination of small probing volume, low concentration of the impurity molecule of interest, and spectral selection is necessary to insure that only one molecule is pumped by the laser beam. In addition, it is important to select a fairly stable host-guest combination and detection technique such that the detected optical signal from one molecule can be observed with sufficient signal-to-noise in a reasonable averaging period.

To proceed from one mole of material (6.02×10^{23} molecules) to only one molecule, many orders of magnitude must be spanned. The spatial coherence available from modern laser sources facilitates the probing of a small volume by providing focal spot sizes on the order of one to a few μm in diameter. It is best to use small sample thicknesses no larger than the Rayleigh range of the focused laser spot (3–10 μm). Thus, in most experiments a small volume of sample on the order of 10–100 μm^3 is probed. This action alone represents an effective reduction of the number of molecules potentially in resonance by some 11 to 12 orders of magnitude, depending on the actual molar volume of the material.

Single-molecule experiments generally utilize samples in which the molecule of interest is present as a dopant or guest impurity in a transparent host matrix. Obviously, then, if the concentration of the guest is sufficiently small, only one molecule of interest will be in the probing volume. For experiments at room temperature where no spectral selection method is available, it is indeed necessary to reduce the concentration of the impurity dramatically to 10^{-12} moles/mole or lower, and to be very sure that no other unwanted impurity in the probed volume is capable of producing a signal that would overwhelm that from the single guest molecule of interest. The experiments in Chapter 2 of this book with near-field optics at room temperature and those in liquids described in Chapter 3 must work precisely in this regime – by extreme reduction of concentration one and only one guest molecule is allowed at a time in the volume pumped by the laser.

However, for high-resolution SMS at low temperatures which is the focus of the first Chapter of this book, such extreme reductions in concentration are not generally required, and samples are usually doped with the guest at concentrations in the range 10^{-7} to 10^{-9} mol/mol. This additional 7 to 9 orders of magnitude reduction in the number of potentially resonant molecules is insufficient to guarantee that only one impurity molecule in the probed volume is in resonance with the laser at a time. The

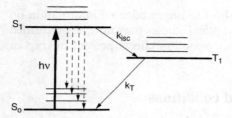

Figure 1. Schematic representation of the electronic energy levels of a molecule showing the ground singlet state S_0, the first excited singlet state S_1, and the lowest triplet state T_1. For each electronic state, several levels in the vibrational progression are shown. Laser excitation pumps the (0–0) transition with energy $h\nu$. The intersystem crossing rate from the singlet manifold to the triplets is k_{isc}, and the triplet decay rate is $k_T = (\tau_T)^{-1}$. Fluorescence emission shown as dotted lines originates from S_1 and terminates on various vibrationally excited levels of S_0.

requisite additional selectivity on the order of a factor of 10^4 or so is provided by spectral selection, which involves carefully selecting the guest and host and using the well-known properties of inhomogeneously broadened absorption lines in solids, to be described next.

1.1.2.2 Spectral selection using zero-phonon lines and inhomogeneous broadening

There is an important physical effect which facilitates the detection and spectroscopy of single molecules in solids at low temperatures, known as inhomogeneous broadening. This effect occurs most clearly when the optical transition pumped by the laser is a purely electronic, zero-phonon line (ZPL), from the lowest vibrational level of the electronic ground state to the lowest vibrational level of the electronically excited state. (We assume that the placement of the guest molecule in the solid effectively hinders rotation of the molecule.) Such a so-called (0–0) transition at energy $h\nu$ shown in Fig. 1 often has a very long lifetime, because the de-excitation to the ground state manifold requires a large number of phonons to be emitted, and such a high-order emission process is improbable. At zero temperature, such a transition could have a lifetime-limited transition width of tens to a few hundred MHz (for an electric-dipole-allowed transition in the visible, even smaller for a partially forbidden transition).

At finite but low temperatures, the width of the ZPL can still be close to the lifetime-limited value. One may wonder why the time-varying perturbations due to the phonons of the solid do not broaden such transitions dramatically. First considering linear coupling to the phonons, it is in fact the extremely high frequency of the phonons compared to the radiative width of the ZPL which places the fluctuations of the optical transition frequency in the motionally-narrowed regime [36], so no broadening of the transition occurs. From the point of view of recoil, the ZPL is often regarded as the optical analog of the Mössbauer line, so that the entire solid sample recoils during optical absorption [37]. As a result of these considerations, a ZPL transition can only dephase and broaden by second-order coupling to the phonons, that is, by phonon scattering (two-phonon processes) [38]. At liquid helium temperatures, few phonons are present to produce phonon scattering, so the homogeneous width γ_H of zero-phonon optical transitions in crystals approaches the lifetime-limited value of some tens of MHz mentioned above. Since the optical transition frequency is near 500 THz, the Q or quality factor of such a narrow transition is very large, near 10^8. In amorphous materials, other low-frequency excitations arising from two-level systems are present, and the homogeneous width is somewhat larger [43], but still far narrower than at high temperatures. It is the extremely high Q of single-molecule lines in solids that leads to exquisite sensitivity to nanoscopic changes – very weak perturbations produced by electric, magnetic, or strain fields in the nanoenvironment can easily produce a detectable shift in the single-molecule absorption.

Now consider what happens for the collection of guest molecules which are located in the probed volume. If the sample were a perfect crystal and all local environments were identical, the optical absorption would be a single narrow Lorentzian[2] line of width γ_H. However, these conditions are seldom met in real solid samples. The optical absorption spectrum that is actually measured for such an assembly is far broader that γ_H. The extreme narrowness of the ZPL for each of the molecules is obscured by a distribution of center frequencies for the various guest molecules, and the resulting overall profile is termed an inhomogeneously broadened line [39, 40] (see Fig. 2). The distribution of resonance frequencies is caused by dislocations, point defects, or random internal electric and strain fields and field gradients in the host material. Such imperfections are generally always present, even in crystals, as long as γ_H is small enough to reveal the inhomogeneous distribution. In the simplest case of inhomogeneous broadening, the overall line profile of width Γ_I is caused by an approximately Gaussian (normal) distribution of center frequencies for the individual absorbers that is broader than the homogeneous lineshape. Inhomogeneous broadening is a universal feature of high-resolution laser spectroscopy of guest molecules in solids [41, 42] and of other cases where ZPLs are probed such as Mössbauer and magnetic resonance spectroscopy. It is for this reason that methods such as spectral hole-burning [44] and coherent transients [45], most often photon echoes, have been utilized to learn about the homogeneous line-width hidden under such inhomogeneous spectral profiles.

Fortunately, the normally troublesome phenomenon of inhomogeneous broadening facilitates the spectral selection of individual molecules for SMS. In effect; the spread of center frequencies means that different guest molecules have different resonance frequencies, so if the total concentration is low enough, one simply uses the tunability of a narrowband laser to select different single molecules. This spectral selection must be done in a region of the frequency spectrum where the number of molecules per γ_H is on the order of or less than one. In general, this may be accomplished in several different ways: (a) by using a sample with a low doping level, (b) by using a sample with a very large Γ_I, or (c) by tuning out into the wings of the inhomogeneous line as shown in the right side of Fig. 2. A useful analogy is provided by

---

[2] The normalized Lorentzian absorption profile centered at ω_o with full-width at half-maximum γ_H is $(\gamma_H/2\pi)/[(\omega - \omega_o)^2 + (\gamma_H/2)^2]$

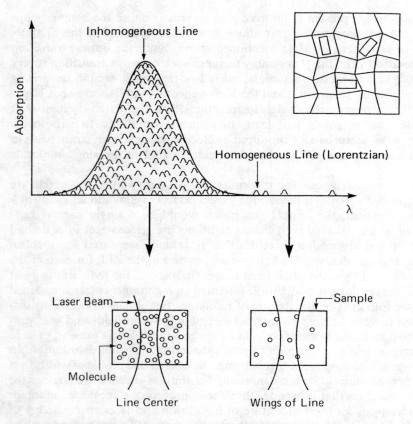

Figure 2. Schematic showing an inhomogeneous line at low temperatures and the general principle of single-molecule detection in solids. The entire line is formed as a superposition of (generally Lorentzian) homogeneous profiles of the individual absorbers, with a distribution of center resonance frequencies caused by random strains and imperfections which can be normally distributed (Gaussian). In the upper right, several dopant molecules are sketched as rectangles with different local environments produced by strains, local electric fields, and other imperfections in the host matrix. The lower part of the figure shows how the number of impurity molecules in resonance in the probed volume can be varied by changing the laser wavelength. The laser line-width is negligible on the scale shown. Although this figure shows that wavelengths in the wings of the line are required to achieve the single-molecule limit, this is not essential if either the concentration of guest molecules is lowered or if the inhomogeneous linewidth is increased.

the problem of tuning in a radio station when one is out in the country where only a few stations can be received. As the receiver is tuned, mostly static is received (no signal) until the exact frequency of a distant station is reached. Similarly, when inhomogeneous broadening causes the different single molecules in the probed volume to have different resonance frequencies, and the molecules are spaced apart by more than γ_H on average, single molecules can be pumped selectively, one at a time, simply by tuning the laser.

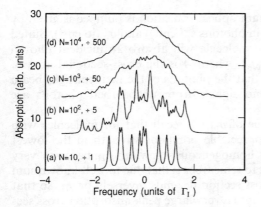

Figure 3. Simulated absorption spectra with different total numbers of absorbers N, using Lorentzian profiles for the individual absorbers and a Gaussian random variable to select center frequencies in the inhomogeneous line. Traces (a) through (d) correspond to N values of 10, 100, 1000, and 10 000, respectively, and the traces have been divided by the factors shown. For clarity, the homogeneous width is taken to be one-tenth of the inhomogeneous standard deviation Γ_I.

To provide a slightly more realistic picture of the inhomogeneous broadening phenomenon, Fig. 3 shows a simulated inhomogeneous lineshape produced by the superposition (summation) of a large number of individual homogeneous Lorentzian absorption profiles. In this simulation, Γ_I/γ_H was chosen to be 10 for simplicity. (In many physical systems, the ratio Γ_I/γ_H is much larger, often in the range 10^3 to 10^6.) In trace (a), only ten centers were superposed to produce the entire "inhomogeneous" line, and the identification of structures corresponding to single molecules is clearly evident. The other traces show the result of successively increasing the number of centers N in the probed volume. The "spectral roughness" called statistical fine structure [46] on the peak of the profile is clearly evident, although it is decreasing in relative magnitude as $1/(\sqrt{N})$. In real experiments, statistical fine structure is often observed before proceeding to the SMS regime, in order to optimize the detection system. Indeed, the early observation of statistical fine structure for pentacene in p-terphenyl crystals [46, 47] provided useful encouragement that the single-molecule regime could eventually be reached. With difficulty, statistical fine structure has also been observed for ions in inorganic materials [48, 49].

1.1.2.3 Peak absorption cross section

A central concept which is crucial to single-molecule spectroscopy in solids is the role of the peak absorption cross section of the guest molecules σ_p. Study of this parameter helps to answer the question: Why is the signal from a single molecule large enough to be detected above background in a reasonable period of time? As will be shown below, the signal-to-noise ratio for SMS depends crucially on maximizing the value of σ_p no matter which technique is used for detection. For focal spots of area greater than or equal to the diffraction limit (approximately λ^2, with λ the optical wavelength), the rate at which the resonant optical transition is pumped is given by the product of the incident photon flux (in photons s^{-1} cm^2)) and σ_p (in cm^2). Stated differently, the probability that a single molecule will absorb an incident photon from the pumping laser beam is just σ_p/A, where A is the cross-sectional area of the focused laser beam. High σ_p means that the photons of the incident light beam are efficiently absorbed and background signals from unabsorbed photons are minimized.

To understand what controls the value of σ_p, we recall that for molecules with weak electron-phonon coupling with appreciable oscillator strength in the lowest purely electronic transition, the optical homogeneous linewidth γ_H becomes very small at low temperatures. The key point to remember is that due to well-known sum rules on optical transitions, the peak cross section depends inversely or γ_H, so that the narrow linewidth of a ZPL translates into a very large peak absorption cross section. For allowed transitions of rigid molecules, the value of σ_p becomes extremely large, approaching the ultimate limit of λ^2.

Although the importance of the peak absorption cross section has been recognized since the first SMS experiment [1], it is useful to review how σ_p can be estimated using sum-rule techniques. The standard integrated absorption sum rule [50–54] may be written

$$S = \int \alpha(\tilde{v})\mathrm{d}\tilde{v} = \left(\frac{\pi e^2}{mc^2}\right)\left(\frac{F}{n}\right)f N_{\text{tot}} = \frac{8\pi^3\tilde{v}_0}{ch(F/n)}\left(\frac{|\vec{\mu}|^2}{3}\right)N_{\text{tot}} \tag{1}$$

where e is the electronic charge, the local field factor is $F = [(n^2 + 2)/3]^2$, n is the index of refraction, c is the speed of light, N_{tot} is the number density of absorbers producing the integrated absorption S (units cm^{-2}), f is the oscillator strength, $\tilde{v}_o$ is the frequency at the center of the band in wavenumbers, and $|\vec{\mu}|$ is the transition dipole moment magnitude. Because the oscillator strength (or dipole moment) is generally independent of temperature, the ratio S/N_{tot} is a constant which may either be evaluated for the homogeneous band at room temperature or for a single Lorentzian profile at low temperature. Applying Eq. 1 to the single-molecule Lorentzian absorption profile of width $\Delta\tilde{v} = (\pi c T_2)^{-1} = \gamma_H/(2\pi c)$ (where T_2 is the dephasing time) and identifying the peak absorption divided by the number density of absorbers producing that absorption as the peak cross section (per center) yields the standard formula [20]

$$\sigma_p = \left(\frac{2\pi e^2 T_2}{mc}\right)\left(\frac{L}{n}\right)f_{\text{ZPL}} = \left(\frac{16\pi^3 T_2\tilde{v}_o}{h}\right)\left(\frac{L}{n}\right)\left(\frac{|\vec{\mu}|^2 \text{ZPL}}{3}\right)$$

$$= 2cT_2\left(\frac{S}{N_{\text{tot}}}\right)_{\text{ZPL}} \tag{2}$$

If the integral to determine S in Eq. 1 is performed for a room-temperature liquid solution, the resulting value of oscillator strength f applies to the entire electronic transition, including all the phonon and vibrational sidebands of the ZPL. To obtain the oscillator strength of the ZPL only, f_{ZPL}, for use in Eq. 2, the value of f must be multiplied by $F_{FC}F_{DW}$, where F_{FC} is the Franck–Condon factor and F_{DW} is the Debye–Waller factor. Similar considerations apply to the dipole moment and integrated absorption strength S. Thus, with measurement of the low temperature linewidth (or T_2) and either the oscillator strength, dipole moment, or the value of S/N_{tot} for the transition, σ_p can be determined. For an equivalent alternative approach based on radiative lifetimes, see Ref. 55.

For molecules like pentacene, perylene, or similar rigid aromatics with a strongly allowed lowest electronic transition (see Section 1.1.3.3 for structures), the net result is that at low temperatures where T_2 is large (some tens of ns), the peak absorption cross section increases to levels as high as 10^{-11} cm^2, approximately 4 000 times the (van der Waals) area of a single molecule! Thus, even though any dimension of a single molecule is much smaller than the optical wavelength, the effective area of the molecule for optical absorption is not nearly so small, and a zero-phonon optical transition of a molecule in a solid can be made to absorb light quite efficiently if the sample is cooled to low temperatures. To give a specific example for the case of pentacene, the low-temperature homogeneous width is 7.8 MHz [56], the measured ZPL dipole moment without orientational averaging is 0.7 Debye [57] where one Debye is defined as 3.33×10^{-30} C m, and thus the value of σ_p is estimated to be near 9×10^{-12} cm^2.

1.1.2.4 Other important requirements for single-molecule spectroscopy

This section briefly describes several additional requirements necessary to insure that the signal from a single molecule dominates over all background signals (for full details see Refs. 5 and 20).

In the case of fluorescence detection, the quantum yield for photon emission per absorption event ϕ_p should clearly be high, as close to unity as possible. Extreme care should be taken to minimize scattering backgrounds which may arise either from Rayleigh scattering, or from Raman scattering from the sample and the substrate.

A further requirement on the absorption properties of the guest molecule stems from the general fact that higher and higher laser power generally produces more and more signal, as long as the optical transition is not saturated. When saturation occurs, further increases in laser power generate more background rather than signal, and this is true for both fluorescence and absorption detection methods. The saturation intensity is maximized when absorbing centers are chosen that do not have strong bottlenecks in the optical pumping cycle (see Fig. 1). In organic molecules, intersystem crossing (ISC) from the singlet states into the triplet states represents a common bottleneck, because both absorption of photons and photon emission cease for a relatively long time equal to the triplet lifetime when ISC occurs. This effect results in premature saturation of the emission rate from the molecule and reduction of the absorption cross section σ_p compared to the case with no bottleneck [58]. For later reference, the saturation intensity I_S for a molecule with a triplet bottleneck may be written [59, 60]

$$I_S = \frac{hv}{2\sigma_P T_1} \left[\frac{1 + (k_{isc}/k_{21})}{1 + (k_{isc}/2k_T)} \right] \qquad (3)$$

where $T_1 = 1/k_{21}$ is the inverse of the rate of direct decay from S_1 to S_0, k_{isc} is the rate of intersystem crossing as shown in Fig. 1, and k_T is the total decay rate from the triplet back to S_0. The factor outside the brackets is the two-level saturation intensity if there were no triplet bottleneck which represents an upper limit for the saturation intensity. Thus, minimizing the triplet bottleneck means small values of k_{isc} and large values of k_T, requirements which may be easily satisfied for rigid, planar aromatic dye molecules.

A final requirement for SMS is the selection of a guest–host couple that allows for photostability of the impurity molecule and weak spectral hole-burning, where by spectral hole-burning we include any fast light-induced change in the resonance frequency of the molecule caused either by frank photochemistry of the molecule or by a photophysical change in the nearby environment [44]. For example, most fluorescence detection schemes with overall photon collection efficiency of 0.1% to 1% require that the quantum efficiency for hole-burning be less than 10^{-6} to 10^{-7}. This is necessary to provide sufficient time averaging of the single-molecule signal before it changes appreciably or moves to another spectral position.

It should be noted that the additional requirements for SMS stated in this section represent a "best-case" in order to produce the highest possible signal-to-noise in a high-resolution experiment. If some loss in signal-to-noise or spectral resolution can be tolerated, these requirements can be weakened accordingly. Specific forms of the SNR for various detection methods will be presented in the next section.

1.1.3 Methods

1.1.3.1 Geometrical configurations for focusing and fluorescence collection

A variety of different configurations have been used for achieving the required focusing of the radiation in a small sample at liquid helium temperatures. If a direct absorption method is used, one need only collect the transmitted light and direct it to the detector. In the case of fluorescence detection, it is necessary to carefully collect the emitted fluorescence photons from the single molecule over as large a solid angle as possible, without collecting unwanted transmitted pumping radiation or scattered light. Because of the large number of experiments utilizing fluorescence detection, all configurations described here refer to the fluorescence method.

One useful experimental setup is shown in Fig. 4(a), the "lens–parabola" configuration. All the components shown can be immersed in superfluid helium. The focal spot from the pumping laser is produced by a small lens of several mm focal length placed directly in the liquid helium. Generally, some provision must be made for adjusting the focus at low temperatures. In one solution to this [58], the lens is mounted on a thin stainless steel plate, which can be flexed by a permanent magnet

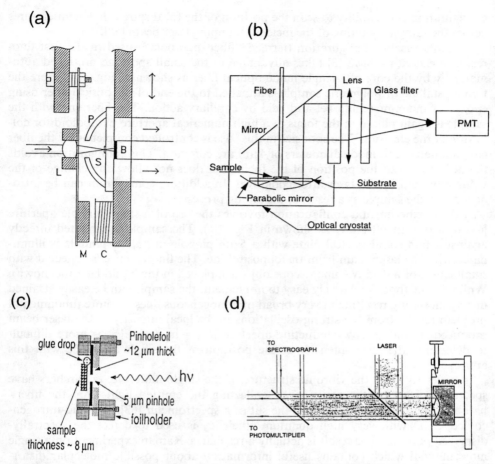

Figure 4. Various experimental arrangements for SMS in fluorescence excitation: (a) lens–paraboloid (after Ref. 58), L – lens, P – paraboloid, S – sample, B – beam block, M – magnet, C – coil electromagnet, (b) fiber–paraboloid (first used in Ref. 3), (c) pinhole (after Ref. 61), (d) paraboloid focus and collection (after Ref. 65).

M/electromagnet C pair to adjust the focal position. The sample is mounted on a transparent substrate, ideally an alkali halide, whose center of symmetry prevents first-order Raman scattering by the substrate. After passage through the sample, the transmitted pumping radiation is blocked by a small beam block. The emitted fluorescence is collected by a paraboloid with numerical aperture near 1.0, and directed out of the cryostat. Using a standard two-element achromatic lens, a spot size of 3–5 μm diameter can be produced. The size of the focal spot is limited by the distortion and aberrations produced by cooling a lens designed for operation at room temperature. With optimized optics able to tolerate liquid helium temperatures, smaller spot sizes should be achievable with this configuration. One feature of this configuration is the ability to scan the position of the focal spot at low temperatures across the sample by tilting of the incident pumping laser beam [58].

A second useful configuration (termed "fiber–parabola") used in the first fluorescence excitation SMS [3] takes advantage of the small spot size produced automatically by the core of a single mode optical fiber as shown in Fig. 4(b). Here the thin crystalline or polymeric sample is attached to the end of the optical fiber using epoxy- or index-matching gel and held by capillary action. The fiber end with the sample is again placed at the focus of a high numerical aperture paraboloid for collection of the emission. The spot size in this case is controlled completely by the fiber core diameter, and mode diameters of 4 µm are common. This configuration avoids the need to adjust the position of the focus, but does not allow any change of the volume probed by the laser after cooldown. In addition, some strain can be introduced into the sample as a result of the gluing process.

A third experimental configuration involves the use of a small pinhole aperture [61] in a thin metal plate, as shown in Fig. 4(c). The sample is mounted directly against a thin stainless steel plate with a 5 µm pinhole in it. The pinhole is illuminated with the laser beam from the opposite side. The fluorescence is collected with small lenses or a 0.85 NA microscope objective placed in the liquid He (not shown). While this method is relatively easy to implement, the sample can be easily strained during mounting resulting in very broad inhomogeneous lines. A more fundamental problem results from the strong oscillations in the local intensity of the laser beam produced by the nearby conducting aperture [62] – in general it is more difficult to determine the laser intensity at the position of the single molecule with this approach.

In order to study the vibronic structure of the single molecule, researchers have spectrally dispersed the emission by collecting the emitted light from the fiber–parabola setup and focusing it on the slit of a spectrometer [63]. A liquid-nitrogen-cooled CCD with very high quantum efficiency is used to detect the spectrally-dispersed photons. The result is actually a resonance Raman experiment on a single molecule [64] which contains useful information about possible molecular distortions produced by the local environment. In a recent enhancement of this technique shown in Fig. 4(d) [65], a high-quality diamond-turned Al paraboloid was used both to focus the emitting radiation, as well as to collect the emission in an epi-fluorescence configuration. The sample was placed at the focus of the paraboloid by flexing a thin metal plate. The focal spot diameter was estimated to be less than 2 µm, and the total sample volume probed as small as 10 µm².

For experiments in which microwave and magnetic fields are required to pump transitions between spin sublevels, a sophisticated enhancement of the lens-parabola design has been described [66] which allows low temperature positioning of the sample along two axes without use of electro- or permanent magnets.

1.1.3.2 Detection techniques

Direct absorption (frequency-modulation spectroscopy with secondary modulation)

The first single-molecule spectra were recorded in the pentacene in *p*-terphenyl system in 1989 using a sophisticated zero-scattering-background absorption technique,

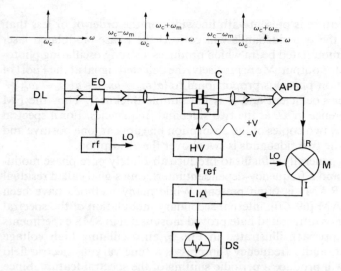

Figure 5. Schematic of frequency-modulation spectroscopy with Stark secondary modulation (FM-Stark). The top of the figure shows the light spectrum at the output of the dye laser (DL), after the electro-optic phase modulator (EO), and after the cryostat (C, from left to right), the arrows indicating the relative amplitudes and phases of the electric light fields. The rf source drives the EO modulator and produces the local oscillator signal (LO) for the mixer M. The output of the avalanche photodiode (APD) drives the R port of the mixer. The simple FM signal is present at the I port of the mixer. For secondary Stark modulation, a high-voltage source (HV) at reference frequency f produces a time-varying electric field across the sample. A lock-in amplifier detects the output of the I port of the mixer, and the resulting signal is averaged on a digital oscilloscope (DS). After Ref. 48.

frequency-modulation (FM) spectroscopy [67, 68], combined with either Stark or ultrasonic modulation of the absorption line [1, 2]. Rather than describe the complete details of the FM technique here, it is more useful to describe the basic characteristics of the method, and the reader may consult Refs. 2 and 5 for more information. Referring to Fig. 5, using an electro-optic modulator (EO), the dye laser (DL) at frequency ω_c is phase-modulated at the local oscillator (LO) radio frequency ω_m which produces two sidebands on the laser carrier at $\omega_c + \omega_m$ and at $\omega_c - \omega_m$ with opposite phase. Any amplitude modulation of the laser beam exiting the sample is detected using a fast photodiode (for example, an APD) and a phase-sensitive radio frequency lock-in (mixer M) driven at ω_m. If no narrow spectral features are present, there is only a dc signal at the detector output, since a perfect phase-modulated laser beam has no amplitude modulation. The background noise level at the I port of the mixer is produced by detector noise and by the laser noise at ω_m. For LO frequencies above about 1 MHz, the limiting noise source can be quantum (shot) noise if no excess noise is introduced by the detector. Generally, the laser power at the detector must be above a certain minimum value for shot noise to dominate over other noise sources.

If a narrow spectral feature is present with linewidth on the order of or less than ω_m, the unbalancing of the two sidebands will convert the phase-modulated laser beam into an amplitude-modulated beam which produces a strong oscillating photo-current at ω_m at the detector output. More precisely, the detected signal at the I port of the mixer (in the absorption phase) is proportional to $[\alpha(\omega_c + \omega_m) - \alpha(\omega_c - \omega_m)]L$ where α is the absorption coefficient and L is the sample thickness. Thus the FM signal measures the difference in αL at the two sideband frequencies. For a spectral feature narrower than ω_m, two copies of the absorption line appear, one positive and one negative, as each of the two sidebands is swept over the absorption.

In actual practice, it is generally difficult to produce absolutely pure phase modulation of the laser beam, and a frequency-dependent interfering signal called residual amplitude modulation (RAM) is often present. While many methods have been proposed to eliminate RAM [69–70], internal secondary modulation of the spectral features by Stark-shifting or ultrasound hate proved most useful in SMS experiments [2]. For the Stark-FM approach illustrated in Fig. 5, an oscillating high voltage (HV) source driven at an audio frequency f impresses a time-varying electric field across the sample. This will produce a periodic shifting of the spectral feature. Since the RAM does not oscillate at frequency f, the RAM may be removed by detection of the mixer I-port output with a lock-in amplifier (LIA) driven at f (for a linear Stark shift) or at $2f$ (for a quadratic Stark shift). The resulting lineshape will appear as the appropriate spectral derivative of the FM signal (Ref. 1).

The signal-to-noise ratio for a single molecule detected by FM spectroscopy has been described in detail in Ref. 5. Since this is an absorption technique, clearly single molecules with higher absorption cross section lead to larger FM signals, and detectors with internal gain such as APD's are helpful in reducing detector noise. In contrast to fluorescence methods, Rayleigh scattering and Raman scattering are unimportant. Only the shot noise of the laser beam contributes to the background, assuming RAM and detector noise are properly controlled.

Fig. 6 shows examples of the optical absorption spectrum from a single molecule of pentacene in p-terphenyl using the FM Stark method. Although this early observation and similar data from the FM ultrasound method served to stimulate much further work, there is one important limitation to the general use of FM methods for SMS. As was shown in the early papers on FM spectroscopy [67, 68], extremely low absorption changes as small as 10^{-7} can be detected in a 1 s averaging time, but only if large laser powers on the order of several mW can be delivered to the detector to reduce the relative size of the shot noise. This presents a problem for SMS in the following way. Since the laser beam must be focused to a small spot, the power in the laser beam must be maintained below the value which would cause power broadening of the single-molecule lineshape. As a result, it is quite difficult to utilize laser powers in the mW range for SMS of allowed transitions at low temperatures – in fact powers below 100 nW are generally required. This is one reason why the SNR of the original data on single molecules of pentacene in p-terphenyl in Fig. 6 was only on the order of 5. (The other reason was the use of relatively thick cleaved samples, which produced a larger number of out-of-focus molecules in the probed volume. This problem has been overcome with much thinner samples in modern experiments.) If either materials with higher saturation intensity or squeezed light beams

Figure 6. The first single-molecule optical spectra, showing use of the FM/Stark technique for pentacene in *p*-terphenyl. (a) Simulation of absorption line with (power-broadened) linewidth of 65 MHz. (b) Simulation of FM spectrum for (a), $\omega_m = 75$ MHz. (c) Simulation of FM/Stark line-shape. (d) single-molecule spectra at 592.423 nm, 512 averages, 8 traces overlaid, bar shows value of $2\omega_m = 150$ MHz. (e) Average of traces in (d) with fit to the in-focus molecule (smooth curve). (f) Signal far off line at 597.514 nm. (g) Traces of SFS at the O_2 line center, 592.186 nm. After Ref. 1.

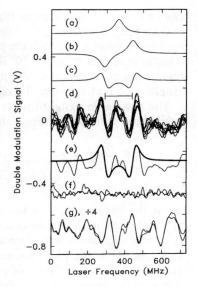

with reduced shot noise become easily available in the future, the utility of the FM method will improve.

Fluorescence excitation spectroscopy

In 1990, Orrit et al. also began experiments on the pentacene in *p*-terphenyl system and demonstrated that fluorescence excitation spectroscopy produces superior signals if the emission is collected efficiently and the scattering sources are minimized [3]. Most subsequent experiments have used this technique, in which a tunable narrowband single-frequency laser is scanned over the absorption profile of the single molecule, and the presence of absorption is detected by measuring the fluorescence emitted. A long-pass filter is used to block the pumping laser light, and the fluorescence shifted to long wavelengths is detected with a photon-counting system, usually a photomultiplier and discriminator. The detected photons generally cover a broad range of wavelengths, because the emission from the ground vibrational level of the electronically excited state terminates on various vibrationally excited (even) levels of the electronic ground state as shown in Fig. 1.

In fluorescence excitation, the detection is background-limited and the shot noise of the probing laser is only important for the signal-to-noise of the spectral feature, not the signal to background. For this reason, it is critical to efficiently collect photons (as with a paraboloid or other high numerical aperture collection system), and to reject the pumping laser radiation. To illustrate suppose a single molecule of pentacene in *p*-terphenyl is probed with 1 mW cm^{-2}, near the onset of saturation of the absorption due to triplet level population. The resulting incident photon flux of 3×10^{15} photons s^{-1} cm^2 will produce about 3×10^4 excitations per second. With a fluorescence quantum yield of 0.8 for pentacene, about 2.4×10^4 emitted photons can be expected. At the same time, 3×10^8 photons/s illuminate the focal spot 3 µm in diameter. Considering that the resonant 0–0 fluorescence from the molecule must be thrown away along with the pumping light, rejection of the pumping radiation by a factor greater than 10^5 to 10^6 is generally required, with minimal attenuation of the fluorescence. This is often accomplished by low-fluorescence glass filters or by holographic notch attenuation filters.

The attainable signal-to-noise ratio (SNR) for single molecule detection in a solid using fluorescence excitation can be approximated by the following expression [71]:

$$\frac{S_1}{(\text{noise})_{\text{rms}}} = \frac{(D\phi_F\sigma_pP_o\tau)/(Ah\nu)}{\sqrt{(D\phi_F\sigma_pP_o\tau)/(Ah\nu) + C_bP_o\tau + N_d\tau}} \tag{4}$$

where the numerator, S_1, is the peak detected fluorescence counts from one molecule in an integration time τ, ϕ_F is the fluorescence quantum yield, σ_p is the peak absorption cross section on resonance as defined above, P_o is the laser power, A is the focal spot area, $h\nu$ is the photon energy, N_d is the dark count rate, and C_b is the background count rate per Watt of excitation power. The factor $D = \eta_Q F_P F_f F_l$ describes the overall efficiency for the detection of emitted photons, where η_Q is the photomultiplier tube quantum efficiency, F_p is the fraction of the total emission solid angle collected by the collection optics, F_f is the fraction of emitted fluorescence which passes through the long pass filter, and F_l is the total transmission of the windows and additional optics along the way to the photomultiplier. The three noise terms in the denominator of Eq. 4 represent shot noise contributions from the emitted fluorescence, background, and dark signals, respectively. For a detailed discussion of the collection efficiency for a single molecule taking into account the dipole radiation pattern, total internal reflection, and the molecular orientation, see Ref. 60.

Assuming the collection efficiency D is maximized, Eq. 4 shows that there are several physical parameters which must be chosen carefully in order to maximize the SNR. First, as stated above, the values of ϕ_F and σ_p should be as large as possible, and the laser spot should be as small as possible. The power P_o cannot be increased arbitrarily because saturation causes the peak absorption cross section to drop from its low-power value σ_o according to [72]

$$\sigma_p \to \sigma_P(I) = \sigma_o/(1 + I/I_S) \tag{5}$$

where I is the laser intensity and I_S is the characteristic saturation intensity. The effect of saturation in general can be seen in both the peak on-resonance emission rate from the molecule $R(I)$ and in the single-molecule linewidth $\Delta\nu(I)$ according to [58]:

$$R(I) = R_\infty\left[\frac{I/I_S}{1 + (I/I_S)}\right] \tag{6}$$

$$\Delta\nu(I) = \Delta\nu(0)[1 + (I/I_S)]^{1/2} \tag{7}$$

Figure 7. Signal-to-noise ratio for fluorescence excitation of pentacene in *p*-terphenyl versus probing laser power and laser beam cross-sectional area. After Ref. 5.

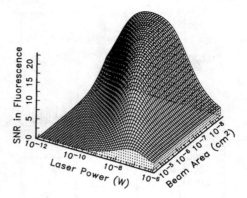

maximum emission rate is given by

$$R_\infty = \frac{(k_{21} + k_{\mathrm{isc}})\phi_{\mathrm{F}}}{2 + (k_{\mathrm{isc}}/k_{\mathrm{T}})} \tag{8}$$

Eqs. 5 and 7 show that the integrated area under the single-molecule peak falls in the strong saturation regime. However, at higher and higher laser power, more and more scattering signal is produced in proportion to the laser power, so the difficulty of detecting a single molecule increases. The dependences of the maximum emission rate and linewidth on laser intensity in Eqs. 6 and 7 have been verified experimentally for individual single molecules [58].

To illustrate graphically the tradeoffs inherent in Eqs. 4 and 5, parameters for the model system pentacene in *p*-terphenyl will be used, for which $\phi_{\mathrm{F}} = 0.78$ [57] and $D \approx 0.01$ in the lens–parabola geometry. Measured values for the background scattering level, dark count rate, and other parameters are given in Ref. 5. Eqs. (4) and (5) then yield a relationship between the SNR, P_{o}, and A. Assuming 1 s integration time, Fig. 7 shows the SNR versus laser power and beam area. It is clear that for a fixed laser spot size, an optimal power exists which maximizes the tradeoff between the saturating fluorescence signal and the linearly (with P_{o}) increasing background signal. For fixed spot size, the SNR at first improves, because the signal increases linearly with laser power and the shot noise from the power-dependent terms in the denominator only grow as the square root of the laser power. As saturation sets in, however, the SNR falls because the signal no longer increases. Another relationship illustrated in Fig. 7 is that the best-case SNR at smaller and smaller beam areas levels off (the flattening of the ridge at small beam area). This is due to the effect of saturation and shot noise – at smaller and smaller areas the power must be reduced eventually to the point where the SNR is controlled by the shot noise of the detected signal (first term in the denominator of Eq. 4). In any case, SNR values on the order of 20 with 1 s averaging time are quite useful for spectroscopic studies. However, more information about dynamical effects can be obtained if the SNR is increased further, which is one continuing challenge to the experimenters in this field. From another point of view, improvements in the SNR would allow probing at lower laser

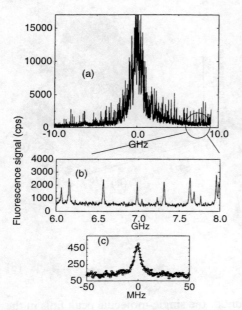

Figure 8. Fluorescence excitation spectra for pentacene in *p*-terphenyl at 1.5 K measured with a tunable dye laser of linewidth 3 MHz. The laser detuning frequency is referenced to the line center at 592.321 nm. (a) Broad scan of the inhomogeneously broadened line; all the sharp features are repeatable structure. (b) Expansion of 2 GHz spectral range showing several single molecules. (c) Low-power scan of a single molecule at 592.407 nm showing the lifetime-limited width of 7.8 MHz and a Lorentzian fit. After Ref. 7.

another point of view, improvements in the SNR would allow probing at lower laser power so that materials with non-optimal photophysics (such as higher hole-burning quantum efficiency) may be studied.

To provide an example of specific experimental spectra using the fluorescence excitation method, we again turn to the model system of pentacene in *p*-terphenyl for simplicity. Fig. 8 shows fluorescence excitation spectra at 1.5 K for a 10 µm thick sublimed crystal of *p*-terphenyl doped with pentacene using the lens–paraboloid setup [7]. The 18 GHz spectrum in Fig. 8(a) (obtained by scanning a 3 MHz linewidth dye laser over the entire inhomogeneous line) contains 20 000 points; to show all the fine structure usually requires several meters of linear space. The structures appearing to be spikes are not noise; all features shown are static and repeatable. Near the certer of the inhomogeneous line, the statistical fine structure (SFS) characteristic of $N > 1$ is observed. It is immediately obvious that the inhomogeneous line is far from Gaussian in shape and that there are tails extending out many standard deviations from the center both to the red and to the blue. Fig. 8(b) shows an expanded region in the wing of the line. Each of the narrow peaks is the absorption profile of a single molecule. The peak heights vary due to the fact that the laser transverse intensity profile is bell-shaped and the molecules are not always located at the center of the laser focal spot. Even though these spectra seem narrow, they are in fact slightly power-broadened by the probing laser.

Upon close examination of an individual single-molecule peak at lower intensity (Fig. 8(c)), the lifetime-limited homogeneous linewidth of 7.8 ± 0.2 MHz can be observed [73]. This linewidth is also termed "quantum-limited", since the optical linewidth has reached the minimum value allowed by the lifetime of the optical excited state. This value is in excellent agreement with previous photon echo measurements using large ensembles of pentacene molecules [56, 57]. Such well-isolated, narrow single-molecule spectra in Fig. 8 are wonderful for the spectroscopist: many detailed spectroscopic studies of the local environment can be performed, because such narrow lines are much more sensitive to local perturbations than broad spectral features.

It is instructive at this point to compare the signal-to-noise ratio for SFS ($N \gg 1$) to that for one single molecule (Eq. 4). Defining N_H as the number of molecules with resonance frequency within one homogeneous width of the laser frequency, and recalling that the SFS signal excursions scale as the square root of the number of molecules in resonance (see Fig. 3),

$$\frac{S_{SFS}}{(noise)_{rms}} = \frac{\sqrt{N_H}(S_1)}{\sqrt{(N_H S_1) + C_b P_0 \tau + N_d \tau}} \approx \sqrt{D\phi_F \left(\frac{\sigma_p}{A}\right)\left(\frac{P_0 \tau}{h\nu}\right)} \qquad (9)$$

where the last approximation assumes that background and dark counts are negligible. In this limit, the SNR is is *independent* of the number of molecules in resonance! Therefore, when detection of SFS has been accomplished, the SNR at that point is a good estimate for the SNR for one single molecule, which degrades only when large background and dark counts are present. It is also interesting that the SNR for SFS scales as the inverse square root of the beam area. Of course, since fluorescence excitation is not a zero-background method, the SFS signal is still a small signal with a relative size $S_{SFS}/S_{TOTAL} = 1/(\sqrt{N_H})$ which must be detected on a large background, thus laser amplitude drifts and low frequency noise must be minimized in order to see the fine structure. (It is precisely this last point which stimulated the original SFS experimenters [46] to utilize FM spectroscopy.)

Single-molecule imaging in frequency and space

With the ability to record high-quality single-molecule absorption lineshapes such as those in Fig. 8, it becomes interesting to acquire spectra as a function of the position of the laser focal spot in the sample. Clearly, a single molecule should also be localized in space as well as absorption frequency. Using the lens–parabola geometry, the laser focal spot can be scanned over a small range in the transverse spatial dimension, and spectra can be obtained at each position. Fig. 9 shows such a three-dimensional "pseudo-image" of single molecules of pentacene in *p*-terphenyl [73]. The *z*-axis of the image is the usual fluorescence excitation signal, the horizontal axis is the laser frequency detuning (300 MHz range), and the axis going into the page is the transverse spatial dimension produced by scanning the laser focal spot (40 μm range). There are three, large, clear single molecule peaks localized in both frequency and position at the center, upper left, and upper right. The resolution of this image in the spatial dimension is clearly limited by the 5 μm diameter laser spot; in fact, the single molecule is actually serving as a highly localized nanoprobe of the laser beam diameter itself. However, in the frequency dimension the features are fully resolved. It is clear that single-molecule peaks are localized in both frequency and space. Extensions of this concept to true diffraction-limited "fluorescence microscopy" using two

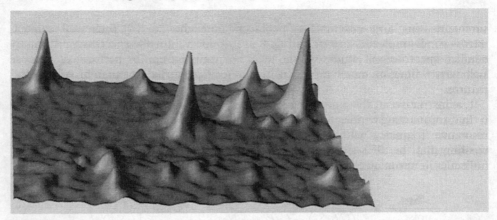

Figure 9. Three-dimensional "images" of single molecules of pentacene in *p*-terphenyl. The measured fluorescence signal (*z*-axis) is shown over a range of 300 MHz in excitation frequency (horizontal axis, center = 592.544 nm) and 40 μm in spatial position (axis into the page).

spatial dimensions will be described in Chapter 1.2, and extensions providing spatial resolution beyond the optical diffraction limit will be described in Chapter 2.

Measurement of spectral trajectories of single molecules

When a new regime is first opened for study, often new physical effects can be observed. In the course of the early SMS studies of pentacene in *p*-terphenyl, an unexpected phenomenon appeared: resonance frequency shifts of individual pentacene molecules in a crystal at 1.5 K [24], called "spectral diffusion" by analogy to similar shifting behavior long postulated for amorphous systems [74]. Here, spectral diffusion means changes in the center (resonance) frequency of a defect due to configurational changes in the nearby host which affect the frequency of the electronic transition via guest–host coupling. In the pentacene in *p*-terphenyl system, two distinct classes of single molecules were identified: class I, which have center frequencies that are stable in time like the three large molecules in Fig. 9, and class II, which showed spontaneous, discontinuous jumps in resonance frequency of 20–60 MHz on a 1–420 s time scale, an example of which is responsible for the distorted single-molecule peak in the center right region of Fig. 9.

Spectral shifts of single-molecule lineshapes are common in many systems, including crystals, polymers, and even polycrystalline Shpol'skiii matrices [75, 76]. Spectral diffusion effects will be described in detail experimentally in Section 1.4, with theoretical analysis in Section 1.5. In this section and in the next, the principal experimental methods for studying this behavior will be described. Again taking pentacene in *p*-terphenyl as an example, Fig. 10(a) shows a sequence of fluorescence excitation spectra of a single molecule taken as fast as allowed by the available SNR. The laser was scanned once every 2.5 s with 0.25 s between scans, and the hopping of this molecule from one resonance frequency to another from time to time is clearly evident.

Figure 10. Examples of single-molecule spectral diffusion for pentacene in *p*-terphenyl at 1.5 K. (a) A series of fluorescence excitation spectra each 2.5 s long spaced by 2.75 s showing discontinuous shifts in resonance frequency, with zero detuning = 592.546 nm. (b) Trend or trajectory of the resonance frequency over a long time scale for the molecule in (a). (c) Resonance frequency trend for a different molecule at 592.582 nm at 1.5 K and at (d) 4.0 K. After Ref. 7.

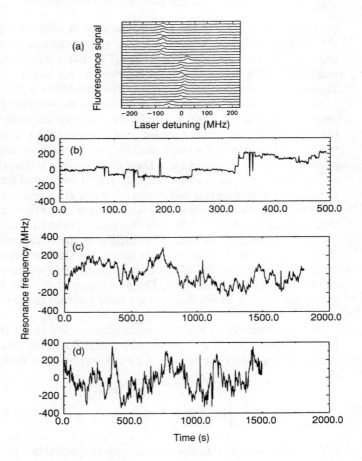

One useful method for studying such behavior is the measurement of the *spectral trajectory* $\omega_o(t)$ [24]. By sequentially acquiring hundreds to thousands of fluorescence excitation spectra and utilizing the power of digital processing to retain a record of the resonance frequency position of each such spectrum, a trajectory or trend of the resonance frequency versus time $\omega_o(t)$ can be obtained as shown in Fig. 10(b) (for the same molecule as Fig. 10(a)). For this molecule, the optical transition energy appears to have a preferred set of values and performs spectral jumps between these values that are discontinuous on the 2.5 s time scale of the measurement. The behavior of another molecule is shown in Fig. 10(c) at 1.5 K and in Fig. 10(d) at 4.0 K. This molecule wanders in frequency space with many smaller jumps, and both the rate and range of spectral diffusion increase with temperature suggesting a phonon-driven process.

The first question which should be asked when such behavior is observed is this: is the effect spontaneous, occurring even in the absence of the probing laser radiation, or is it light-driven, i.e., produced by the probing laser itself. To answer this question, it is usually necessary to observe the spectral shifting behavior as a function of the probing laser power. In the case of the type II pentacene molecules in p-terphenyl. the spectral diffusion appeared to be a spontaneous process rather than a light-induced spectral hole-burning effect [73], but other materials have shown light-induced shifting behavior [25, 75, 80], which may be regarded as the single-molecule analog of the nonphotochemical spectral hole-burning process [77].

Since the optical absorption for pentacene in p-terphenyl is highly polarized [78] and the peak signal from the molecule does not decrease when the spectral jumps occur, it is unlikely that the molecule is changing orientation in the lattice. Since the resonance frequency of a single molecule in a solid is extremely sensitive to the local strain field, the conclusion from these observations is that the spectral jumps are due to internal dynamics of some configurational degrees of freedom in the surrounding lattice. The situation is analogous to that for amorphous systems, which are postulated to contain a multiplicity of local configurations that can be modeled by a collection of double-well potentials (the two-level system or TLS model [43]). The dynamics results from phonon-assisted tunneling or thermally activated barrier crossing in these potential wells. One possible source for the tunneling states [58] could be discrete torsional librations of the central phenyl ring of the nearby p-terphenyl molecules about the molecular axis. The p-terphenyl molecules in a domain wall between two twins or near lattice defects may have lowered barriers to such central-ring tunneling motions.

Spectral trajectories contain much information about the stochastic behavior of the single molecule. If a simple measure of the average time scale of spectral shifts is required, it is useful to calculate the autocorrelation of the spectral trajectory, $C_\omega(\tau)$, given by

$$C_\omega(\tau) = \int \omega_o(t)\omega_o(t+\tau)\mathrm{d}t \tag{10}$$

which is the Fourier transform of the power spectral density of the frequency fluctuations. More complex measures can be computed, such as the survival probability for the resonance frequency to stay at its initial value, and examples of this type of analysis are presented in Section 1.5.

Although the measurement of the spectral trajectory in principle contains all the dynamical information about the system, there are practical limitations to the information that can be obtained. The principal shortcoming of the spectral trajectory measurement results from the time required to scan the absorption line with sufficient SNR to determine the resonance frequency. In many systems, this minimum scanning time is limited by the photon emission rate to times on the order of 10–100 ms, which means that dynamical behavior on faster time scales is not adequately represented. A partial solution to this problem is provided by direct time correlation measurements on the emission signal itself to be discussed in the next section. Nevertheless, the direct observations of the dynamics of a nanoenvironment of a single molecule by the spectral trajectory method have sparked fascinating new theoretical studies of the underlying microscopic mechanism [34, 35, 79], described in detail in Section 1.5. It is worth noting that such spectral trajectories cannot be obtained when a large ensemble of molecules is in resonance. The individual jumps are gen-

Figure 11. Schematic of the temporal behavior of photon emission from a single molecule showing bunching on the scale of the triplet lifetime (upper half) and antibunching on the scale of the inverse of the Rabi frequency (lower half).

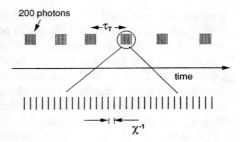

erally uncorrelated, thus the behavior of an ensemble-averaged quantity such as a spectral hole would only be a broadening and smearing of the line.

Time correlation of single-molecule emission signal

A useful route to single-molecule information on shorter time scales lies in the actual photon emission process itself. Let us assume that the pumping laser frequency is held fixed in resonance with the single molecule. The stream of photons emitted by the molecule contains information about the system encoded in the arrival times of the individual photons. Fig. 11 schematically shows the time-domain behavior of the photon stream for a single molecule with a dark triplet state, here taken to be pentacene for definiteness (see Fig. 1). While cycling through the singlet states $S_o \rightarrow S_1 \rightarrow S_o$, photons are emitted until intersystem crossing occurs. Since the triplet yield is 0.5% [57], 200 photons are emitted on average before a dark period which has an average length equal to the triplet lifetime, τ_T. This causes "bunching" of the emitted photons as shown in the upper half of the figure. Even though the actual length of time in the triplet state is a random variable with an exponential distribution, the bunching can be easily detected by measuring the autocorrelation of the photon emission signal $S(t)$, which is defined by

$$C_S(\tau) = \int S(t)S(t+\tau)dt \tag{11}$$

Autocorrelation analysis has long been recognized as a useful method for statistical study of stochastic dynamical processes which may be obscured by noise [81, 82]. By definition, the autocorrelation measures the similarity (overlap) between the function and a copy of the function delayed in time by a lag time τ. Because the noise component of the signal is uncorrelated, its contribution to $C_s(\tau)$ decays quickly, leaving information about the (average) time-domain correlations of $S(t)$. In practice, when photon counting is used, a commercial digital correlator is employed to measure $C_s(\tau)$ by keeping track of the arrival times of many photon pairs. Modern correlators can do this over a huge logarithmic time scale covering many decades in time, a feature that is very useful for studying the dispersive dynamics characteristic of amorphous systems.

The decay in the autocorrelation of the emitted photons for pentacene in *p*-terphenyl due to the triplet bottleneck was first reported by Orrit et al. [3], and this method has been used to measure the changes in the triplet yield and triplet lifetime from molecule to molecule [26] which occur as a result of distortions of the molecule by the local nanoenvironment. For single molecules in polymers with complex dynamics driven by TLS's in the host matrix, correlation measurements are also quite useful [27, 83]. In this case, the amplitude fluctuations in the single-molecule fluorescence signal resulting from shifts of the resonance frequency can produce a characteristic fall-off in the autocorrelation which yields information about the TLS flipping rate distribution. Detailed examples of the use of this method to study spectral diffusion will be presented in Section 1.4.

Correlation measurements can extract information about the single molecule on much shorter time scales (down to the µs range) than the frequency scans described in the previous section. At the same time, however, the dynamical process must be stationary, that is, the dynamics must not change during the relatively long time (many s) needed to record enough photon arrivals to generate a valid autocorrelation. In addition, since the laser frequency is held fixed, when the molecule hops out of resonance with the laser, all information about the new resonance frequency of the single molecule is lost.

Turning now to the nanosecond time regime (lower half of Fig. 11), the emitted photons from a single molecule can provide still more useful information. On the time scale of the excited state lifetime, the statistics of photon emission from a single quantum system are expected [84] to show photon antibunching, which means that the photons "space themselves out in time", that is, the probability for two photons to arrive at the detector at the same time is small. This is a uniquely quantum-mechanical effect, which was first observed for Na atoms in a low-density beam [85]. Antibunching is fundamentally measured by computing the second-order correlation of the electric field $g^{(2)}(\tau)$ (which is simply the normalized form of the intensity–intensity correlation function $C_s(\tau)$), which shows a drop below the uncorrelated value of unity when antibunching is present [86]. For a single molecule, antibunching is easy to understand as follows. After photon emission, the molecule is definitely in the ground state and cannot emit a second photon immediately. A time on the order of the inverse of the Rabi frequency χ^{-1} must elapse before the probability of emission of a second photon is appreciable. At sufficiently high laser intensity, Rabi oscillations can be observed as the laser coherently drives the single molecule into and out of the excited state before emission occurs.

In actual practice, in order to overcome time limitations caused by the dead time of photomultipliers, two identical detectors are used to measure the distribution of time delays between the arrival times of consecutive pairs of emitted photons as in the classical Hanbury–Brown–Twiss experiment [86], which is described in more detail in Section 1.2. The expected antibunching in single-molecule emission was first observed in the author's laboratory for pentacene in *p*-terphenyl [28], demonstrating that quantum optics experiments can be performed in solids and for molecules for the first time. Of course, if more than one molecule is emitting, the antibunching effect as well as the bunching effect both quickly disappear since the various resonant molecules emit independently. The observation of high-contrast antibunching is strong proof that the spectral features are indeed those of single molecules. Careful

Figure 12. Structures of some of the molecules which have been studied by SMS. (a) Pentacene, (b) perylene, (c) terrylene, (d) tetra-(*t*-butyl)-terrylene (TBT), (e) diphenyloctatetraene (DPOT), (f) 7.8,15.16-dibenzoterrylene (DBT), and (g) 2.3,8.9-dibenzanthanthrene (DBATT).

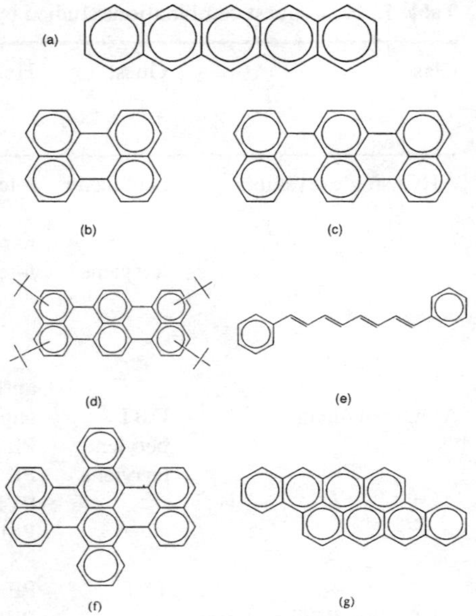

study of the shape of the correlation function in the ns time regime has been used to determine both T_1 and T_2 for a single terrylene molecule in *p*-terphenyl [87].

1.1.3.3 Materials systems and structures

This section briefly lists the materials systems in which high-resolution, low-temperature SMS studies have been performed, to be described in more detail in the remainder of Part I of this book. To date, the guest impurity molecules have been selected exclusively from the class of rigid conjugated hydrocarbons with specific cases shown in Fig. 12: (a) pentacene, (b) perylene, (c) terrylene, (d) tetra-(*t*-butyl)-terrylene (TBT), (e) diphenyloctatetraene (DPOT), (f) 7.8,15.16-dibenzoterrylene (DBT), and (g) 2.3,8.9-dibenzanthanthrene (DBATT). These molecules have strong singlet–singlet absorption, excellent emission properties, and weak triplet bottlenecks. They also feature the weak Franck–Condon distortion necessary to guarantee a strong (0–0) electronic transition. In most cases, workers have concentrated on the fairly large aromatic hydrocarbons (AHCs) (a)–(d), (f), (g) in order to place the lowest electronic transition in the mid-visible. This allows standard tunable single-frequency dye lasers to be utilized for pumping the transition. The one exception, DPOT (e), represents a special case which was pumped either with two-photon excitation in the near IR [88], or with doubled light at 444 nm from the output of a cw Ti–sapphire laser.

The choice of the host material is generally dictated by the need to maintain a weak electron–phonon coupling and to prevent high-efficiency spectral hole-burning. The latter requirement has so far prevented hydrogen-bonded matrices from being suc-

Table 1. Host–guest combinations studied by single-molecule spectroscopy[a].

Class	Guest	Host	ZPL Wavelength (nm)	Linewidth (MHz)	Refs.
AHC: single crystals	pentacene	p-terphenyl	592.32 (O_1), 592.18 (O_2)	7.8 (O_1)	[1, 3, 24, 58]
		naphthalene	602.8	29	[89]
	terrylene	p-terphenyl	580.4 (X_1), 578.5 (X_2), 578.3 (X_3), 577.9 (X_4)	48.1 (X_2)	[90]
		anthracene	–	50–500	[91]
AHC: polymers	DBT	naphthalene	757.7	25–35	[92]
	perylene	PE	∼442	52–142	[25]
	terrylene	PE	569	60–150	[22]
		PVB	562	200–2000	[91]
		PMMA	557	200–2200	[91]
		PS	566	600–4000	[91]
	TBT	PE	–	50–320	[93]
AHC: Shpol'skii matrices		PIB	567.6	40–370	[93]
	terrylene	hexadecane	571.9	40	[75, 76]
	perylene	nonane	∼440	–	[94]
	DPOT	tetradecane	444 (1-photon)	30	[88]
	DBATT	hexadecane	589.1	12–30	[95]

[a] For abbreviations, see text.

cessfully utilized for high-resolution, low-temperature SMS. The actual host–guest combinations fall into three categories: AHCs in crystals, AHCs in polymers, and AHCs in Shpol'skii matrices. Table 1 lists a variety of host–guest combinations that have been studied, along with the position of the ZPL, the single-molecule linewidth (FWHM, generally at ∼2K), and leading references. The favorite single crystal hosts have been p-terphenyl, naphthalene, and anthracene, all of which can be sublimed to form clear micron-thick samples. Generally, single molecules in these hosts are relatively stable, with occasional spontaneous spectral diffusion driven by defects in the host crystal (see Sections 1.4 and 1.5). The polymer hosts have been selected from poly(ethylene) (PE), poly(*i*-butylene) (PIB), poly(vinylbutyral) (PVB), poly(styrene) (PS), and poly(methyl methacrylate) (PMMA). These materials are characterized by highly-dispersive spectral shifting phenomena, both spontaneous and light-driven. In Table 1, a range of linewidths indicates a distribution, and the original references should be consulted for more detail on the shape of this distribution. Finally, various Shpol'skii matrices such as frozen hexadecane, nonane, and tetradecane have also been examined. These convenient polycrystalline hosts yield relatively stable single-molecule spectra, with a slow, light-driven spectral shifting observed in several cases.

To date, the total number of systems which have been studied by high resolution

SMS techniques is approximately 16. Work is in progress on new materials, such as other AHCs and additional Shpol'skii matrices. It is to be expected that some laser dye molecules in appropriate hosts will also have the strong fluorescence, weak triplet bottlenecks, and near-absence of spectral hole-burning required for SMS studies.

1.1.4 Summary and outlook

Application of the concepts described in this chapter have formed the basis for many of the fascinating experiments which have been performed with single molecules in condensed matter, the description of which occupies the remainder of Chapter 1.

The attainment of SMS in solids opens up a new frontier of single-absorber experiments in which the measured properties of the absorbing center are not averaged over many "equivalent" absorbers. The significance of such experiments is fourfold. First, the properties of a single absorber are measured without ensemble averaging, which means that tests of specific theoretical models are much stronger. Second, the sensitivity to specific properties of the nanoenvironment such as the local phonon modes and the true local fields is extremely high. This means for example that the identity of the mysterious two-level systems in amorphous materials may finally be determined. Third, it provides a window into the spectral hole-burning process on a molecule by molecule basis. Thus, the exact local coupling through which optical pumping of a single molecule gives rise to changes in the nanoenvironment which shift the resonance frequency may be studied. Fourth, this regime is essentially unexplored, which means that surprises and unexpected physical effects can occur (such as the observation of spectral diffusion in a crystal).

While as a general technique high-resolution SMS is not applicable to all molecular impurities, it can be applied to the large number of absorbing molecules (and perhaps ions) in solids that have zero-phonon transitions, reasonable absorption strength, and efficient fluorescence. The detectability of the resulting single-center signal, which ultimately depends upon the specific sample and weak or absent spectral hole-burning, must be evaluated in each case. SMS signals should be observable at higher and higher temperatures if the concentration and background are both reduced sufficiently. One successful method for doing this at room temperature has been to use near-field excitation to reduce the scattering volume and increase the single-center signal, to be described in Chapter 2.

One experimental technique that has only recently been utilized for SMS is two-photon fluorescence excitation. In this approach, a long-wavelength source is used to pump the 0–0 transition via two-photon absorption, and the emission is collected as usual. Since two-photon cross sections are much smaller than one-photon cross sections for an electric dipole-allowed transition, high pumping power in the 100 mW to 1 W range is required. This difficulty is offset by the ease with which the pump radiation is separated from the emission. First results for Rhodamine B in water at room temperature have recently been reported [96] using a high-peak-power pulsed laser. At low temperature, the narrowing of the optical absorption yields a huge increase in the two-photon cross section, and pumping with a cw laser has recently produced two-photon optical spectra of single molecules of diphenyloctatetraene in *n*-tetradecane [88].

Other fascinating future experiments may be contemplated based on the principles and methods presented here. Detailed study of the spectral diffusion process in crystals and polymers will help to eventually identify the actual microscopic nature of the two-level systems. The door is open to true photochemical experiments on single absorbers, quantum optics, and even the possibility of optical storage using single molecules [7]. Future efforts to increase the number of probe-host couples which allow SMS will lead to an even larger array of novel observations.

One experiment that is now possible would be to use the emission from a single molecule as a light source of sub-nm dimensions for near-field optical microscopy [97]. Of course, this would involve the technical difficulty of placement of the single emitting molecule at the end of a pulled fiber tip or pipette. In all cases, improvements in SNR would be expected to open up a new level of nanoscopic detail and possibly new applications. Because this field is relatively new, the possibilities are only limited at present by the imagination and the persistence of the experimenter and the continuing scientific interest in the properties of single quantum systems in solids.

References

[1] W. E. Moerner and L. Kador, *Phys. Rev. Lett.* **62**, 2535, (1989).
[2] L. Kador, D. E. Horne, and W. E. Moerner, *J. Phys. Chem.* **94**, 1237, (1990).
[3] M. Orrit and J. Bernard, *Phys. Rev. Lett.* **65**, 2716, (1990).
[4] W. E. Moerner, *New J. Chem.* **15**, 199, (1991).
[5] W. E. Moerner and Th. Basché, *Angew. Chem.* **105**, 537 (1993); *Angew. Chemie Int. Ed. Engl.* **32**, 457, (1993).
[6] M. Orrit, J. Bernard, and R. Personov, *J. Phys. Chem.* **97**, 10256, (1993).
[7] W. E. Moerner, *Science* **265**, 46, (1994).
[8] L. Kador, *Phys. Stat. Sol. (b)* **189**, 11, (1995).
[9] W. E. Moerner, in *Atomic Physics 14*, AIP Conf. Proc. **323**, D. J. Wineland, C. E. Wieman, and S. J. Smith, eds. (AIP Press, New York, 1995), pp. 467–486.
[10] W. M. Itano, J. C. Bergquist, and D. J. Wineland, *Science* **237**, 612, (1987) and references therein.
[11] F. Diedrich, J. Krause, G. Rempe, M. O. Scully, and H. Walther, *IEEE J. Quant. Elect.* **24**, 1314, (1988) and references therein.
[12] H. Dehmelt, W. Paul, and N. F. Ramsey, *Rev. Mod. Phys.* **62**, 525, (1990).
[13] G. Binnig, H. Rohrer, *Rev. Mod. Phys.* **59**, 615, (1987).
[14] G. Binnig, C. F. Quate, and C. Gerber, *Phys. Rev. Lett.* **56**, 930, (1986).
[15] D. Rugar and P. K. Hansma, *Phys. Today* (October 1990), pp. 23–30.
[16] E. Betzig and R. J. Chichester, *Science* **262**, 1422, (1993).
[17] W. P. Ambrose, P. M. Goodwin, J. C. Martin, and R. A. Keller, *Phys. Rev. Lett.* **72**, 160, (1994).
[18] W. E. Moerner, T. Plakhotnik, T. Irngartinger, U. P. Wild, D. Pohl, and B. Hecht, *Phys. Rev. Lett.* **73**, 2764, (1994).
[19] For progress toward the single-ion limit, see S. M. Jaffe, M. L. Jones, and W. M. Yen, *J. Lumin.* **60–61**, 417, (1994).
[20] W. E. Moerner, *J. Lumin.* **58**, 161, (1994).
[21] U. P. Wild, F. Güttler, M. Pirotta, and A. Renn, *Chem. Phys. Lett.* **193**, 451, (1992).
[22] M. Orrit, J. Bernard, A. Zumbusch, and R. I. Personov, *Chem. Phys. Lett.* **196**, 595, (1992).
[23] F. Güttler, T. Irngartinger, T. Plakhotnik, A. Renn, and U. P. Wild, *Chem. Phys. Lett.* **217**, 393, (1994)

[24] W. P. Ambrose and W. E. Moerner, *Nature* **349**, 225, (1991).

[25] Th. Basché and W. E. Moerner, *Nature* **355**, 335, (1992).

[26] J. Bernard, L. Fleury, H. Talon, and M. Orrit, *J. Chem. Phys.* **98**, 850, (1993).

[27] A. Zumbusch, L. Fleury, R. Brown, J. Bernard, and M. Orrit, *Phys. Rev. Lett.* **70**, 3584, (1993).

[28] Th. Basché, W. E. Moerner, M. Orrit, and H. Talon, *Phys. Rev. Lett.* **69**, 1516, (1992).

[29] Ph. Tamarat, B. Lounis, J. Bernard, M. Orrit, S. Kummer, R. Kettner, S. Mais, and Th. Basché, *Phys. Rev. Lett.* **75**, 1514, (1995).

[30] P. Tchénio, A. B. Myers, and W. E. Moerner, *J. Phys. Chem.* **97**, 2491, (1993).

[31] P. Tchénio, A. B. Myers, and W. E. Moerner, *Chem. Phys. Lett.* **213**, 325, (1993).

[32] J. Köhler, J. A. J. M. Disselhorst, M. C. J. M. Donckers, E. J. J. Groenen, J. Schmidt, and W. E. Moerner, *Nature* **363**, 242, (1993).

[33] J. Wrachtrup, C. von Borczyskowski, J. Bernard, M. Orrit, and R. Brown, *Nature* **363**, 244, (1993).

[34] P. D. Reilly and J. L. Skinner, *Phys. Rev. Lett.* **71**, 4257, (1993).

[35] P. D. Reilly and J. L. Skinner, *J. Chem. Phys.* **102**, 1540, (1995).

[36] R. Kubo, *Adv. Chem. Phys.* **15**, 101, (1969).

[37] R. M. Silsbee, *Phys. Rev.* **128**, 1726, (1962).

[38] D. E. McCumber and M. D. Sturge, *J. Appl. Phys.* **34**, 1682, (1963).

[39] A. M. Stoneham, *Rev. Mod. Phys.* **41**, 82, (1969).

[40] K. K. Rebane, *Impurity Spectra of Solids* (Plenum, New York, 1970), p. 99.

[41] D. A. Wiersma, *Adv. Chem. Phys.* **47**, 421, (1981).

[42] See *Laser Spectroscopy of Solids*, Springer Topics in Applied Physics **49**, W. M. Yen and P. M. Selzer, Eds. (Springer, Berlin, 1981).

[43] See *Amorphous Solids: Low-Temperature Properties*, Topics in Current Physics **24**, W. A. Phillips, Ed. (Springer, Berlin, 1981).

[44] See *Persistent Spectral Hole-Burning: Science and Applications*, Topics in Current Physics **44**, W. E. Moerner, Ed. (Springer, Berlin. Heidelberg, 1988).

[45] L. Allen and J. H. Eberly, *Optical Resonance anal Two-Level Atoms*, (Wiley, New York, 1975).

[46] W. E. Moerner and T. P. Carter, *Phys. Rev. Lett.* **59**, 2705, (1987).

[47] T. P. Carter, M. Manavi, and W. E. Moerner, *J. Chem. Phys.* **89**, 1768, (1988).

[48] T. P. Carter, D. E. Horne, and W. E. Moerner, *Chem. Phys. Lett.* **151**, 102, (1988).

[49] W. S. Brocklesby, B. Golding, and J. R. Simpson, *Phys. Rev. Lett.* **63**, 1833, (1989).

[50] A. J. Sievers, *Phys. Rev. Lett.* **13**, 310, (1964).

[51] R. C. Hilborn, *Am. J. Phys.* **50**, 982, (1982).

[52] W. E. Moerner, A. R. Chraplyvy, and A. J. Sievers, *Phys. Rev. B* **28**, 7244, (1983).

[53] W. E. Moerner, M. Gehrtz, and A. L. Huston, *J. Phys. Chem.* **88**, 6459, (1984).

[54] W. E. Moerner, F. M. Schellenberg, G. C. Bjorklund, P. Kaipa, and F. Lüty, *Phys. Rev. B* **32**, 1270, (1985).

[55] K. K. Rebane and I. Rebane, *J. Lumin.* **56**, 39, (1993).

[56] F. G. Patterson, H. W. H. Lee, W. L. Wilson, and M. D. Fayer, *Chem. Phys.* **84**, 51, (1984).

[57] H. de Vries and D. A. Wiersma, *J. Chem. Phys.* **70**, 5807, (1979).

[58] W. P. Ambrose, Th. Basché, and W. E. Moerner, *J. Chem. Phys.* **95**, 7150, (1991).

[59] H. de Vries and D. A. Wiersma, *J. Chem. Phys.* **72**, 1851, (1980).

[60] T. Plakhotnik, W. E. Moerner, V. Palm, and U. P. Wild, *Opt. Commun.* **114**, 83, (1995).

[61] M. Pirotta, F. Güttler, H. Gygax, A. Renn, J. Sepiol, and U. P. Wild, *Chem. Phys. Lett.* **208**, 379, (1993).

[62] A. J. Campillo, J. E. Pearson, S. L. Shapiro, and N. J. Terrell, Jr., *Appl. Phys. Lett.* **23**, 85, (1973).

[63] P. Tchénio, A. B. Myers, and W. E. Moerner, *J. Phys. Chem.* **97**, 2491, (1993).

[64] A. B. Myers, P. Tchénio, M. Zgierski, and W. E. Moerner, *J. Phys. Chem.* **98**, 10377, (1994).

[65] L. Fleury, Ph. Tamarat, B. Lounis, J. Bernard, and M. Orrit, *Chem. Phys. Lett.* **236**, 87, (1995).

[66] H. van der Meer, J. A. J. M. Disselhorst, J. Köhler, A. C. J. Brouwer, E. J. J. Groenen, and J. Schmidt, *Rev. Sci. Instrum.* **66**, 4853, (1995).

[67] G. C. Bjorklund, *Opt. Lett.* **5**, 15, (1980).

[68] G. C. Bjorklund, M. D. Levenson, W. Lenth, and C. Ortiz, *Appl. Phys. B* **32**, 145, (1983).

[69] E. A. Whittaker, M. Gehrtz, and G. C. Bjorklund, *J. Opt. Soc. Am. B* **2**, 1320, (1985).

[70] M. Gehrtz, G. C. Bjorklund, and E. A. Whittaker, *J. Opt. Soc. Am. B* **2**, 1510, (1985).
[71] Th. Basché, W. P. Ambrose, W. E. Moerner, *J. Opt. Soc. Am. B* **9**, 829, (1992).
[72] See for example, P. N. Butcher, D. Cotter, *The Elements of Nonlinear Optics*. Cambridge Univ. Press, Cambridge, 1990, p. 180.
[73] W. E. Moerner and W. P. Ambrose, *Phys. Rev. Lett.* **66**, 1376, (1991).
[74] J. Friedrich and D. Haarer, in *Optical Spectroscopy of Glasses*, I. Zschokke, Ed. (Reidel, Dordrecht, 1986), p. 149.
[75] T. Plakhotnik, W. E. Moerner, T. Irngartinger, and U. P. Wild, *Chimia* **48**, 31, (1994).
[76] W. E. Moerner, T. Plakhotnik, T. Irngartinger, M. Croci, V. Palm, and U. P. Wild, *J. Phys. Chem.* **98**, 7382, (1994).
[77] J. M. Hayes, R. P. Stout, and G. J. Small, *J. Chem. Phys.* **74**, 4266, (1981).
[78] F. Güttler, J. Sepiol, T. Plakhotnik, A. Mitterdorfer, A. Renn, and U. P. Wild, *J. Lumin.* **56**, 29, (1993).
[79] G. Zumofen and J. Klafter, Chem. *Phys. Lett.* **219**, 303, (1994).
[80] P. Tchénio, A. B. Myers, and W. E. Moerner, *J. Lumin.* **56**, 1 (1993).
[81] E. L. Elson and D. Magde, *Biopolymers* **13**, 1, (1974).
[82] D. L. Magde, E. L. Elson, and W. W. Webb, *Biopolymers* **13**, 29, (1974).
[83] L. Fleury, A. Zumbusch, M. Orrit, R. Brown, and J. Bernard, *J. Lumin.* **56**, 15, (1993).
[84] H. J. Carmichael and D. F. Walls, *J. Phys. B* **9**, L43, (1976).
[85] H. J. Kimble, M. Dagenais, and L. Mandel, *Phys. Rev. Lett.* **39**, 691, (1977).
[86] See R. Loudon, *The Quantum Theory of Light*, 2nd Ed. (Clarendon, Oxford, 1983), p. 226–249.
[87] S. Kummer, S. Mais, and Th. Basché, *J. Phys. Chem.* **99**, 17078, (1995).
[88] T. Plakhotnik, D. Walser, M. Pirotta, A. Renn, and U. P. Wild, *Science* **271**, 1703, (1996).
[89] S. Kummer, C. Bräuchle, and Th. Basché, Mol. Cryst. Liq. Cryst. **283**, 255, (1996).
[90] S. Kummer, Th. Basché, and C. Bräuchle, *Chem. Phys. Lett.* **229**, 309, (1994); Chem. Phys. Lett. *232*, 414 (1995).
[91] B. Kozankiewicz, J. Bernard, and M. Orrit, *J. Chem. Phys.* **101**, 9377, (1994).
[92] F. Jelezko, Ph. Tamarat, B. Lounis, and M. Orrit, in press.
[93] R. Kettner, J. Tittel, Th. Basché, and C. Bräuchle, *J. Phys. Chem.* **98**, 6671, (1994).
[94] M. Pirotta, A. Renn, and U. P. Wild (unpublished).
[95] A.-M. Boiron, B. Lounis, and M. Orrit, in press.
[96] J. Mertz, C. Xu, and W. W. Webb, *Opt. Lett.* **20**, 2532, (1995).
[97] K. Lieberman, S. Harush, A. Lewis, and R. Kopelman, *Science* **247**, 59, (1990).

1.2 Excitation and Emission Spectroscopy and Quantum Optical Measurements

T. Basché, S. Kummer, C. Bräuchle

1.2.1 Introduction

The first successful experiments in single molecule spectroscopy (SMS) at low temperatures have been performed on pentacene molecules substitutionally doped into the lattice of a *p*-terphenyl host crystal [1, 2]. This classic system, sometimes called "organic ruby", was already investigated in great detail in numerous experiments with large ensembles of molecules [3, 4] from which its favourable photophysical properties were well known. The absence of spontaneous or photoinduced frequency jumps for the majority of the single pentacene molecules in carefully prepared *p*-terphenyl crystals made this material an ideal candidate for demonstrating numerous effects for the first time in SMS. The probability of environment induced frequency jumps of single molecules increases with the degree of disorder of the host matrix. It is therefore high in amorphous materials and lowest in well defined single crystals. In this chapter we will mainly concentrate on single absorbers that are not subject to the influence of locally evolving structures giving rise to spectral diffusion but where the interactions with the host matrix are mainly due to phonon-induced fluctuations around some equilibrium configuration. The stability of the single molecule excitation lines allows one to perform measurements that require high excitation intensities, long accumulation times or those where frequency jumps would only be a nuisance.

The organization of this chapter is as follows. In Section 1.2.2 we introduce the basic properties of the single molecule excitation lineshape which include its temperature dependence and saturation behaviour. The experimental investigation of the latter two effects is particularly simple because once a single molecule resonance line is isolated no inhomogeneity that complicates the analysis of the data is left. For the proper interpretation of saturation studies, the experimental geometry and simple polarization effects have to be taken into account. The theoretical description used in this section ensues from the knowledge gained during the last decades in the optical spectroscopy of mixed molecular crystals or doped solids in general. According to well established theories of line broadening in solids, the temperature dependence of the optical linewidth can be interpreted by coupling of low frequency modes (librations) to the electronic transition. Saturation effects are described by the steady-state solutions of the optical Bloch equations.

In Section 1.2.3 we demonstrate that vibrationally resolved fluorescence emission spectra of single molecules can be recorded with high signal-to-noise ratio. After description of the instrumentation and observables in single molecule fluorescence spectroscopy we discuss the experimental results for some specific systems. It will be shown that the vibrational structure in the emission spectra can reveal details about the variety of local environments for different single absorbers and isotopic substitution.

The last section of this chapter is devoted to a description of quantum optical experiments which are unique in the sense that their observation requires the investigation of a single quantum system or at least only a small number of them. Quantum jumps between electronic levels fall within this category. Individual singlet–triplet quantum jumps have been observed directly through the large fluctuations of the fluorescence intensity of a single terrylene molecule in a *p*-terphenyl host. The fluorescence intensity autocorrelation function which can be easily recorded over eleven orders of magnitude in time (i.e. nanoseconds – 100 seconds) describes the joint probability of detecting photon pairs separated by time τ. Neglecting the effects of spectral diffusion, the correlation function of individual organic dye molecules contains at least three meaningful time regimes: photon antibunching due to the vanishing joint probability for the emission of two photons at the same time causes a correlation dip at short times which is followed by a coherent transient (Rabi oscillations) when high pumping intensity is used. At long times the correlation function decays exponentially due to the above mentioned singlet–triplet quantum jumps (photon bunching). It will be shown that the fluorescence autocorrelation function allows the determination of fluorescence lifetimes, pure optical dephasing times and population and depopulation times of the triplet state. It is a very powerful technique to perform time resolved measurements on single molecules under continuous wave excitation. Finally, we describe the response of a single molecule to two laser fields providing the basis for experiments of the pump-probe type.

Most of the results presented in the following sections were obtained with single pentacene or terrylene molecules in *p*-terphenyl (see Chapter 1.1) and we briefly want to mention some basic spectroscopic properties of these systems. Both exhibit four distinct electronic origins in absorption and emission. For pentacene in *p*-terphenyl there is general agreement that the pentacene dopant molecules substitutionally occupy the four inequivalent lattice sites of the low temperature triclinic phase of *p*-terphenyl [5] leading to four electronic origins, $O_1 - O_4$ (see Chapter 1.3). In the case of terrylene the appearance of four distinct electronic origins which were named $X_1 - X_4$ [6] suggests the same substitution pattern. The difference in size between terrylene and *p*-terphenyl, however, is substantial and at the moment the 1:1 substitution scheme remains an assumption to be proved. In the remainder of this chapter we will always refer to the above site assignment when we talk about spectroscopic properties of single (pentacene or terrylene) molecules in different electronic origins.

1.2.2 Single molecule optical excitation lineshape

1.2.2.1 Basic properties

The optical absorption or excitation lineshape of dopant molecules in crystalline matrices at low temperatures has been investigated both experimentally and theo-

Figure 1. Zero-phonon line (ZPL) and phonon-side band (PSB) of the electronic excitation of a dopant molecule in a solid host at low temperatures.

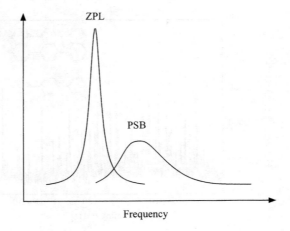

retically in great detail over the last decades [7, 8]. This knowledge paved the road to SMS and forms the basis for the understanding of the single molecule optical excitation lineshape. We will briefly summarize some basic results regarding the optical lineshape in organic mixed crystals. In general, the lineshape is composed of two components, (Fig. 1), the zero-phonon line (ZPL) and the phonon sideband (PSB) [9]. The ZPL describes a transition with no net creation or destruction of phonons or other low-frequency excitations. Linear electron–phonon coupling, where the coupling strength is given by the displacement of the excited state potential surface with regard to the ground state potential surface, gives rise to phonon sidebands of the electronic transition in absorption and emission. Quadratic electron-phonon coupling which will be discussed in Section 1.2.2.2 leads to a temperature dependent broadening of the ZPL by phonon scattering.

As was highlighted in Section 1.1, the appearance of intense and narrow ZPLs forms the basis for single molecule spectroscopy with high signal-to-noise ratio. This immediately limits the number of systems to those which possess very weak electron phonon coupling and hence the oscillator strength of the transition is concentrated in the ZPL. In Fig. 2 the fluorescence excitation spectra of single terrylene molecules doped at very low concentration into a *p*-terphenyl host crystal are shown [6]. Each line corresponds to the purely electronic ZPL of the lowest energy transition $(S_0 \rightarrow S_1)$ of a single absorber. Single molecule excitation spectra of vibronic ZPLs have not been reported so far and their observation will remain an experimental challenge because the lifetime of vibronic levels of a large molecule in a solid is very short (picoseconds) and thus the vibronic ZPLs would be very broad.

As mentioned above, the observation of strong single molecule ZPLs requires weak (linear) electron–phonon coupling. The latter coupling strength, however, is not zero and the reader may wonder why no PSBs are observable in spectra as in Fig. 2. The width of a typical PSB is several tens of wavenumbers and the distance between the maximum of the ZPL and its PSB amounts to a similar number. The typical width of a ZPL, which lies between 10^{-4} and 10^{-3} cm^{-1}, is orders of magnitude smaller and therefore the PSB will not show up in the typical scan range of a single molecule excitation spectrum (Note: The lineshape in Fig. 1 is only sche-

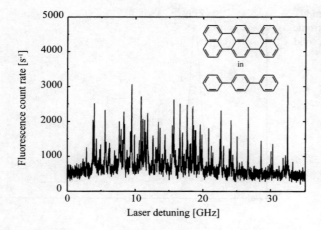

Figure 2. Fluorescence excitation spectrum at 1.4 K of terrylene in a p-terphenyl single crystal in the spectral region of site X_1 ($\sim$ 580.4 nm). The narrow features are the excitation lines of single terrylene molecules. The laser was scanned in 32 seconds over the displayed frequency range at an intensity of 25 mW/cm² (from Ref. 6).

matic.). Additionally, even when scanning over a large frequency range, the intensity of the PSB is very weak and hence its spectrum quite difficult to monitor.

The single molecule ZPLs in Fig. 2 can be very well described by Lorentzians. This fact also holds true for single molecule studies in pentacene/p-terphenyl [10] and terrylene/hexadecane [11]. This result can be understood fairly easy taking into account that in the single molecule experiment all inhomogeneous contributions have been removed. In general, the homogeneous absorption lineshape $I(v - v_0)$ of an electric dipole oscillator is given by the Fourier transform of the autocorrelation function of the transition dipole moment [12]:

$$I(v - v_0) \propto \int dt \exp[i(v - v_0)t]\langle\mu(t)\mu^*(0)\rangle \tag{1}$$

where μ is the transition dipole moment and $v - v_0$ is the detuning of the laser frequency from the transition frequency of the chromophore. The brackets denote averaging over the bath states (phonons). For an exponential decay of the dipole correlation function, it can be shown that the lineshape function $I(v - v_0)$ is given by a Lorentzian. The normalized Lorentzian lineshape function for a ZPL can be written as:

$$I(v - v_0) = \frac{1}{4\pi^2} \frac{\Delta v_{\text{hom}}}{(v - v_0)^2 + (\Delta v_{\text{hom}}/2)^2} \tag{2}$$

where Δv_{hom} is the full width at half maximum (FWHM) of the homogeneously broadened Lorentzian lineshape. The decay of the dipole correlation function is governed by the two time constants T_1 and T_2^* which are related to the homogeneous optical linewidth by the following equation:

$$\Delta v_{\text{hom}} = \frac{1}{\pi T_2} = \frac{1}{2\pi T_1} + \frac{1}{\pi T_2^*} \tag{3}$$

T_2 denotes the dephasing time of the optical transition, T_1 the lifetime of the excited state (fluorescence lifetime) and T_2^* the pure dephasing time. At low temperatures T_1 is essentially independent on temperature while T_2^* shows a strong dependence on temperature. The actual value of T_2^* at a given temperature depends on the excitation of low frequency modes (phonons, librations) that couple to the electronic transition of the chromophore. In crystalline matrices at low temperatures ($T \leq 2\,\mathrm{K}$) T_2^* approaches infinity as host phonons and local modes are essentially quenched and the linewidth is solely determined by the lifetime contribution.

For single pentacene molecules in *p*-terphenyl it was actually shown that at $T = 1.6\,\mathrm{K}$ and low exciting intensities (no saturation broadening, see Section 1.2.2.3), the measured linewidth ($\Delta v_{\mathrm{hom}} = 7.8 \pm 0.2\,\mathrm{MHz}$) [10] corresponds to the lifetime-limited value as given by the fluorescene decay time. The average value of the linewidth ($\Delta v_{\mathrm{hom}} = 40 \pm 10\,\mathrm{MHz}$, $T = 1.4\,\mathrm{K}$) of single terrylene molecules in *p*-terphenyl was found to be close to the lifetime-limited width of 42 MHz as computed from the fluorescence lifetime of terrylene in polyethylene [11]. In this case, however, there is a clear dispersion of the values of Δv_{hom} which is not caused by saturation effects and which can be tentatively attributed to a dispersion of the fluorescence lifetimes. The picture may be quite different for impurities in amorphous matrices where fast changes in the local structure (two-level tunneling transitions) add an additional contribution to the linewidth. Therefore, even at temperatures as low as 1.4 K the linewidth of single absorbers in amorphous matrices is appreciably broadened as compared to the lifetime-limit and varies strongly from molecule to molecule [13–15].

The information that can be obtained from the single molecule excitation (absorption) line in a crystalline solid at low temperatures relates to the shape, width and position of the line. The lineshape was found to be Lorentzian in all investigations so far in agreement with theory and earlier experiments in bulk samples. Because T_2^* (Eq. 3) is strongly temperature dependent, the linewidth increases with temperature and this dependence will be discussed in the next section. The actual frequency position of a single molecule excitation line is determined by the local environment of the molecule in the crystal. The inhomogeneous broadening caused by nearby defects and strains [16–18] allows us to identify individual molecules with absorption frequencies distributed around an average value. One ultimate goal of SMS is to gain knowledge about the microscopic nature of the local environments around the absorbers and their influence on the actual absorption frequency of a specific molecule. A means to study this problem is to measure controlled (reversible) shifts of the absorption frequency induced by external fields (electric fields [19, 20], pressure [21]). Temperature induced lineshifts which will be presented in the next section are mainly dominated by dynamical interactions with the phonon bath.

1.2.2.2 Temperature dependence of the optical linewidth and lineshift

As mentioned before, an increase in temperature activates low frequency modes of the matrix (phonons) or the molecule (librations) which lead to a broadening and a frequency shift of the homogeneous absorption line. In experiments with ensembles of molecules the homogeneous optical linewidth and its temperature dependence

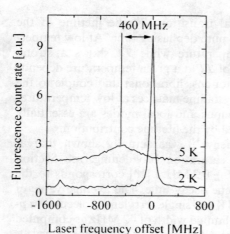

Figure 3. Fluorescence excitation spectra of a single terrylene molecule in *p*-terphenyl (site X_2) at 2 and 5 K. Note the broadening and the shift of the optical line at the higher temperature.

can be extracted in the presence of the ubiquitous inhomogeneous broadening of electronic transitions in low temperature solids by line-narrowing techniques in the time domain as well as in the frequency domain. The former set of experiments comprises coherent optical transients as free induction decay, optical nutation and photon echoes [4, 22], while spectral hole-burning [23] and fluorescence line-narrowing [24] mark the frequency domain techniques. The experimental situation becomes more difficult when extracting the temperature dependent shift of the optical line. Clearly, such information cannot be obtained from coherent optical transients and a technique such as hole-burning must be employed. In many organic mixed crystals, however, hole burning is inefficient or absent and in this case the shift has to be extracted from the temperature dependent position of an absorption line [25, 26] the width of which is inhomogeneously broadened at low temperatures. Then the problem at temperatures below 10 K is to measure shifts that are typically much smaller than the inhomogeneous linewidth.

In contrast, temperature dependent measurements of the optical excitation lineshape of a single molecule immediately give the homogeneous linewidth as well as the lineshift and in this sense SMS can be regarded as the ultimate line-narrowing technique in which no inhomogeneity is left due to ensemble averaging. An example of the optical lineshape of a single terrylene molecule in *p*-terphenyl at two different temperatures [27] is shown in Fig. 3. When raising the temperature, the line broadens dramatically and in this case shifts to the red of the low temperature frequency position. From Fig. 3 the importance of low temperatures for achieving the strongest possible fluorescence signals for single molecules in organic crystals can be immediately deduced. Here, however, the authors were explicitly interested in the functional form of the temperature dependence and hence had to sacrifice SNR.

The first temperature dependent study of the optical linewidth of individual molecules was performed on single pentacene molecules doped into *p*-terphenyl [10]. As can be seen in Fig. 4, below 4 K the optical linewidth remains essentially constant at the lifetime-limited value of 7.8 MHz. Above 4 K, temperature dependent dephasing processes contribute to the linewidth. In previous photon echo experiments, the

Figure 4. Temperature broadening of the optical line width of single pentacene molecules in *p*-terphenyl (site O_1). The various symbols refer to different molecules. The solid line is a fit of Eq. 4 to the data (from Ref. 10).

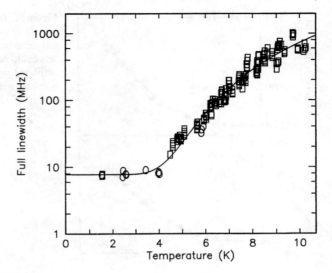

temperature dependence of the echo decay time for pentacene in *p*-terphenyl (O_1) was found to obey a model of dephasing via a single pentacene librational mode [26]. The data in Fig. 4 were fit with the following expression for dephasing [10]:

$$\Delta\nu_{\text{FWHM}}(T) = \frac{1}{\pi T_2} = \frac{1}{2\pi T_1} + \frac{1}{\pi T_2'(\infty)}\exp(-\Delta E/kT) \qquad (4)$$

where k denotes Boltzmann's constant. $\Delta E = 27 \pm 0.7\,\text{cm}^{-1}$ is the librational energy and the pure dephasing prefactor is $1/\pi T_2'(\infty) = 33.5 \pm 4$ GHz or $T_2'(\infty) = 9.5$ ps. Both values are in agreement with previous results from ensemble measurements [26, 28].

In recent experiments the temperature dependence of the optical linewidth $\Delta\nu_{\text{hom}}$ and lineshift $\Delta\nu_s$ was investigated for single terrylene molecules in *p*-terphenyl in site X_2 [27]. In this case no independent information from time or frequency domain studies on bulk samples was available and SMS served as the technique to obtain the pertinent parameters. The temperature dependence of the linewidth as well as the lineshift could be fitted by single exponentials of a functional form similar to that in Eq. 4. For the data displayed in Fig. 5a librational energy of $\Delta E = 17 \pm 2\,\text{cm}^{-1}$ for both the width and the shift of this specific molecule was found. This result can be interpreted in terms of the well-known "exchange model" for optical dephasing [30] which predicts that both the frequency shift and the increase in linewidth are exponentially activated. Similar results were obtained for free base porphin in an *n*-octane crystal by means of spectral hole-burning [31]. Within the "exchange model" the preexponential factors for the width and shift have a different meaning than in Eq. 4 and contain the lifetime and frequency of the local mode in the electronic ground and excited state [8]. As no independent spectroscopic information of these parameters was available, a detailed interpretation of the pre-exponential factors (which only served as fitting parameters) was not yet possible.

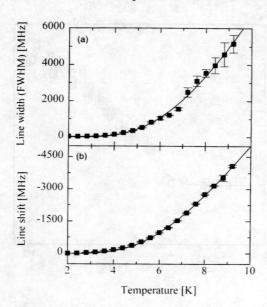

Figure 5. Temperature dependence of the optical linewidth and lineshift for a single terrylene molecule in *p*-terphenyl. Both data sets could be approximated by fitting an activated process (Eq. 4) to the data. The width as well as the shift yield an activation energy of $17 \pm 2\,\mathrm{cm}^{-1}$.

It is interesting to see that for both dopants (pentacene and terrylene) in *p*-terphenyl optical dephasing seems to be induced by a librational mode. This model is strengthened by the observation of different activation energies for the two chromophores which gives further evidence that the low frequency excitations are localized at the impurities.

The temperature dependence of the linewidth was consistently found to be activated with an energy of $18 \pm 3\,\mathrm{cm}^{-1}$ for different single terrylene molecules in *p*-terphenyl. With regard to the lineshift, however, in the authors group molecules were found where the lineshifted to the blue at low temperatures (2–3.5 K) and to the red at higher temperatures ($T > 4\,\mathrm{K}$). While those data so far could not be analyzed quantitatively, they may be explained qualitatively as follows. At low temperatures, where the contribution of the activated process is weak, the crystal lattice surrounding the molecule may expand and shift the line to the blue. Thermal expansion is expected to lead to a blue shift, because, to look at the phenomenon from the opposite direction, when a molecule possesing an electronic π–π^* transition is embedded in a matrix, the solvent shift with regard to the vacuum transition frequency is always to the red. At higher temperatures the activated process may dominate and shift the single molecule line to the red. While this explanation has to be regarded as very tentative, these observations demonstrate the sensitivity of the temperature dependent lineshift to the local environment which may change from molecule to molecule.

1.2.2.3 Saturation behaviour

At high intensities of the exciting laser the linewidth and the fluorescence emission rate of a single molecule are subject to saturation effects. Before we outline the theo-

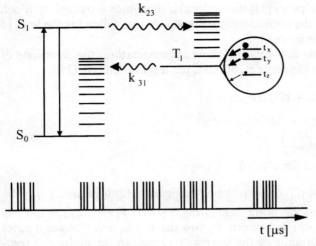

Figure 6. Energy level scheme for a typical aromatic hydrocarbon. S_0 denotes the electronic ground state, S_1 the first excited singlet state and T_1 the first excited triplet state. The triplet state is actually split into three sublevels by magnetic dipolar interaction of the triplet electrons (zero-field splitting). The dots and arrows denote the approximate populations and lifetimes of the sublevels for a typical, planar aromatic hydrocarbon. The lower panel shows schematically the time distribution of fluorescence photons (photoelectric pulses) for a single emitter undergoing singlet-triplet transitions leading to photon bunching.

retical description of these phenomena, we will briefly discuss the transitions between the electronic three-level structure of an organic dye molecule (Fig. 6) [32]. In a typical experiment the molecule is irradiated with laser light in resonance with the $S_0 \rightarrow S_1$ zero-phonon transition. From S_1 the molecule may fluoresce and decay to the ground state or, with very low probability, undergo intersystem crossing (ISC) from S_1 to T_1. Due to the spin-forbidden nature of singlet–triplet transitions, the lifetime of T_1 (between micro- and milliseconds) is orders of magnitude longer than of S_1 (≈ 10 nanoseconds). For large aromatic hydrocarbons absorbing in the red part of the optical spectrum, ISC is predominantly a nonradiative process which occurs between isoenergetic levels of the singlet and triplet manifold. As both singlet states are different in energy from T_1, vibrational quanta are always involved in ISC. The incoherent transition rates k_{23} and k_{31} determine the ISC dynamics because the vibrational relaxation of a large molecule in a solid occurs in picoseconds. At the extreme right of Fig. 6 the zero-field splitting of the triplet state [33] by the magnetic dipole interaction of the triplet electrons is shown (for details see Section 1.6). For the present we will neglect this splitting and treat T_1 as a single level.

In principle the photophysical dynamics of a molecule under coherent excitation has two components. In the coherent motion driven by the laser field, the pseudospin representing the density matrix of the electronic system performs a Rabi precession between the singlet ground and excited states. Additionally, incoherent transitions through coupling to the vacuum field modes or to intramolecular vibrational modes will cause radiative and non-radiative population transfer between the electronic levels. In the theoretical description [34] the molecular dynamics is treated by Bloch equations for the elements of the corresponding density matrix, following the lines of earlier work by deVries and Wiersma [35].

Applying the electric dipole and rotating-wave approximation, the equations of motion for the density matrix elements σ_{ij} of the three-level system are [34],

$$\dot{\sigma}_{11} = k_{21}\sigma_{22} + k_{31}\sigma_{33} + i\Omega/2(\sigma_{21} - \sigma_{12})$$

$$\dot{\sigma}_{22} = -(k_{21} + k_{23})\sigma_{22} + i\Omega/2(\sigma_{12} - \sigma_{21})$$

$$\dot{\sigma}_{33} = k_{23}\sigma_{22} - k_{31}\sigma_{33}$$

$$\dot{\sigma}_{12} = -i\Omega/2\sigma_{11} + i\Omega/2\sigma_{22} + (i\Delta - \Gamma_2)\sigma_{12}$$

with $\sigma_{11} + \sigma_{22} + \sigma_{33} = 1$. $\Omega = |\vec{\mu}_{12}\vec{E}_L|/\hbar$ is the on-resonance Rabi-frequency, $\vec{\mu}_{12}$ the transition dipole moment and $\vec{E}_L$ is the laser field. $\Gamma_2 = 1/T_2 = 1/2T_1 + 1/T_2^*$ is the dephasing rate of the S_0–S_1 coherence, k_{ij} are the incoherent transition rates depicted in Fig. 6. Δ is the detuning of the laser with respect to the molecular resonance frequency.

The Bloch equations (Eq. 5) can be solved under different conditions. The transient solution yields an expression for σ_{22} (t), the time-dependent population of the excited singlet state S_1. It will be discussed in detail in Section 1.2.4.3 in connection with the fluorescence intensity autocorrelation function. Here we are interested in the steady state solution ($\dot{\sigma}_{11} = \dot{\sigma}_{22} = \dot{\sigma}_{33} = \dot{\sigma}_{12} = 0$) which allows to compute the lineshape and saturation effects. A detailed description of the steady state solution for a three level system can be found in [35]. From those the appropriate equations for the intensity dependence of the excitation linewidth $\Delta\nu_{FWHM}$ (FWHM: full width at half maximum) and the fluorescence emission rate R for a single absorber can be easily derived [10]:

$$\Delta\nu_{FWHM}(I) = \Delta\nu_{FWHM}(0)(1 + I/I_S)^{1/2} \tag{6}$$

$$R(I) = \sigma_{22}\frac{1}{\tau_F}\phi_F = R_\infty \frac{I/I_S}{1 + I/I_S} \tag{7}$$

I denotes the intensity, I_S the saturation intensity, $\tau_F = (k_{21} + k_{23})^{-1}$ the fluorescence lifetime and ϕ_F the fluorescence quantum yield. The expression for the saturation intensity appearing in the density-matrix treatment (MKS units) is:

$$I_s = \frac{\varepsilon_0 c\hbar^2(k_{21} + \sum_i k_{23}^i)}{|\vec{\mu}_{21}|^2(2 + A)T_2}, \tag{8}$$

where $A = \sum_i k_{23}^i/k_{31}^i$.

The fully saturated emission rate is given by:

$$R_\infty = \frac{(k_{21} + \sum_i k_{23}^i)\phi_F}{2 + A} \tag{9}$$

Figure 7. Saturation behaviour of single pentacene molecules in *p*-terphenyl (site O_1; $T = 1.4$ K). The (a) fluorescence excitation line width (FWHM) and (b) the fluorescence emission rate are shown as a function of laser intensity. The solid lines are fits to the data as described in the text. The intensity is the free-space peak intensity at the molecule without local-field corrections (from Ref. 10).

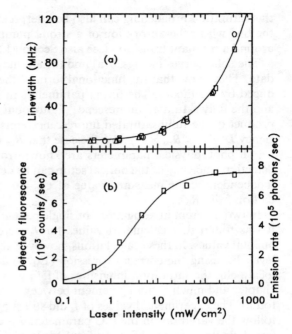

The sums over the triplet sublevels i in Eqs. 8 and 9 take into account that the triplet state is not given by a single level (see Fig. 6), a fact that was neglected in the Bloch equations (Eq. 5). Therefore, in the accurate description, we have to treat a five-level rather than a three-level system. We now want to give some examples for the application of Eqs. 6 and 7 to intensity dependent single molecule data. In Fig. 7 the intensity saturation behaviour for several individual pentacene molecules in the O_1 site of the *p*-terphenyl crystal at 1.5 K is shown [10]. Fig. 7(a) shows single molecule linewidth broadening and Fig. 7(b) shows peak fluorescence saturation with excitation intensity. The free space peak intensity, I, of the laser was computed from the measured power and the laser spot size at the location of the molecule. To accurately determine the laser spot size, the laser beam was translated spatially across the face of the crystal to measure the spatial width of a single molecule excitation peak. This is a direct replica of the Gaussian intensity profile of the laser beam in the crystal, as probed with the fluorescence emitted by the molecule. This careful determination of the laser spot size can be achieved with the lens focusing setup described in Section 1.1 which principally allows the alignment of the single molecule (i.e. its absorption cross section) with respect to the center of the Gaussian laser beam.

At low exciting intensities the linewidths (Fig. 7(a)) reach a value of 7.3 ± 0.8 MHz in agreement with the lifetime-limited value, and the fluorescence emission rate increases almost linearly. In the high-intensity limit, the peak fluorescence emission rate saturates at $7.2 \pm 0.7 \times 10^5$ photons/s (Fig. 7(b)). As the peak emission rate saturates, the emission rate in the wings of the excitation spectrum continues to increase with intensity, and the linewidth broadens as shown in Fig. 7(a). Actually, this saturation broadening can also be interpreted as an optical Stark-broadening of the line where the absorption of a strong pump field versus frequency is measured around a resonant transition (see also Section 1.2.4.4).

The solid curves in Figs. 7(a) and (b) are fits of Eqs. 6 and 7, respectively, to the data. They show that the functional form of the saturation behaviour is well reproduced by the theory. The fitting parameters in Eq. 7 are the saturation intensity I_s and the fully saturated fluorescence photocount rate (C_∞). The latter is given by the product of the fully saturated fluorescence emission rate R_∞ and the detection efficiency $D(C_\infty = R_\infty \times D)$. For the case that R_∞ can be reliably calculated by Eq. 9 – i.e. all photophysical parameters are known from independent measurements – the detection efficiency of the optical setup can be determined. Conversely, if D is known independently, the measured value of C_∞ can be compared to the theoretical prediction from R_∞.

In two different investigations of single pentacene molecules in p-terphenyl [10, 36] it was found that calculated values for I_s were all much smaller than the experimental values. In these calculations, however, the triplet state was treated as a single level. By using the correct expression (Eq. 8) and approximate corrections for the local field, the saturation intensities of [36] were recalculated by the authors of this chapter and much better agreement between experiment and theory than in [36] was found. The experimental values of I_s did still show quite a large scatter which did not follow the variations in the ISC parameters k_{23} and k_{31} which were found to be different up to a factor of three from molecule to molecule [36] (see also 1.2.4.3). Therefore, this distribution of saturation intensities may arise due to differences in local fields and variations in the orientation of the molecular transition dipole moments with respect to the electric field of the exciting laser.

For single terrylene molecules embedded in hexadecane [11] or p-terphenyl [6] the I_s values were found to be much larger than those calculated by Eq. 8. Recently, it was shown that for the analysis of saturation data the collection efficiency and the orientation of the molecular dipole moment with respect to the light polarization and the collecting optics have to be properly included [37]. For the special case of a single terrylene molecule in hexadecane it was found that for the experimental setup used the angle between the emission dipole moment and the optical axis of the light collection paraboloid was $16 \pm 6°$. Assuming that the absorption and emission dipoles are parallel, the polarization vector of the exciting laser light ($\vec{k}$ was parallel to the optical axis) was almost perpendicular to the transition dipole moment of the molecule. As I_s depends on the projection of $\vec{\mu}_{12}$ on the electric field, the measured value of the saturation intensity $(I_{s(\text{exp})} = 1\,\text{W}\,\text{cm}^{-2})$, derived by fitting Eq. 7 to the corresponding data, was much higher than the value calculated from Eq. 8 $(I_{s(\text{calc.})} \sim 210\,\text{m}\,\text{W}\,\text{cm}^{-2})$. After applying their careful analysis Wild et al. [37] estimated the true saturation intensity of terrylene in hexadecane (for dipole moment parallel to the electric field of the light) to be about $80 \pm 30\,\text{mW}\,\text{cm}^{-2}$. The huge discrepancy between experimental values of I_s (10–$20\,\text{W}\,\text{cm}^{-2}$) [6] and the theoretical prediction for terrylene molecules in p-terphenyl is also mainly due to the almost perpendicular orientation of the molecular transition dipole moment with respect to the laser polarization [29].

Figure 8. Sample illumination and light collection geometry to record vibrationally resolved fluorescence spectra of a single molecule. The sample is attached to the optical fiber. S: sample, P: paraboloidal collection mirror, M: flat mirror, BS: beam splitter, F: long pass filter, L: lens, PMT: photomultiplier tube, CCD: charge-coupled device detector (after Ref. 38).

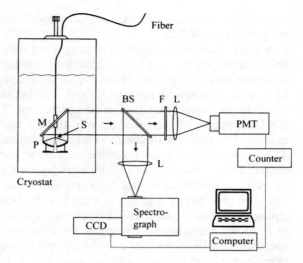

1.2.3 Fluorescence spectroscopy

1.2.3.1 Basic instrumentation

Most single molecule experiments in solids at low temperatures described in this book have been performed using fluorescence excitation techniques in which the pure electronic absorption spectrum is monitored by detection of the total Stokes shifted fluorescence as a function of the excitation frequency. There is, however, also the possibility to excite the molecule with a fixed frequency in the maximum of its absorption line and disperse the emitted fluorescence light by a monochromator. In this fashion the vibrationally resolved fluorescence spectrum of a single molecule can be recorded.

In Fig. 8 a typical illumination and detection setup for single molecule fluorescence spectroscopy as reported by Moerner and coworkers [38] is given. The illumination setup is similar to that described in Section 1.1. The light from a single frequency laser is brought into the cryostat through a single mode polarization-preserving optical fiber having a core diameter of about 4μm. The sample, either a thin crystal or a polymer film, is attached to the end of the fiber with a minimal amount of adhesive. The tip of the fiber is positioned within the cryostat at the focus of a paraboloidal reflector with a numerical aperture of 1.0. The laser light transmitted through the sample passes through a small hole in the vertex of the reflector. The fluorescence is collected and collimated by the paraboloid, reflected 90° by a flat mirror, directed out of the cryostat, and divided by a 50/50 beamsplitter. The transmitted light travels through a long-pass filter to a photomultiplier with photon counting electronics as a detector of "total fluorescence". This arm can also be used to record the fluorescence excitation spectrum of the specific molecule investigated when scanning the exciting laser. The light reflected by the beamsplitter is imaged with a lens onto the entrance slit of a spectrograph and detected with a charge-coupled device (CCD).

An important prerequisite to achieve good signal-to-noise ratio in single molecule fluorescence spectroscopy is to efficiently image the faint emission from a single molecule onto the entrance slit of the spectrograph. This requires a carefully aligned, high quality parabolic mirror [39] and – as far as possible – avoidance of any elements in the optical path that cause aberrations. The spectral resolution depends on the focal length of the spectrograph, the slit width and the groove density of the grating. The detector of choice is a back-illuminated, liquid nitrogen-cooled CCD. These devices have quantum efficiencies between 50 and 75% and virtually no dark current ($\approx 1\,e^-$ per pixel per hour) allowing long accumulation times without build-up of excessive instrumental background. When the single molecule fluorescence signal is very strong and stable even a Peltier-cooled CCD is sufficient to record emission spectra. Besides, a CCD is also advantageous compared to a photomultiplier as it allows multichannel detection, i.e. a reasonably large part of the fluorescence spectrum can be covered in a single recording without moving the grating. Specific experimental conditions will be given in Section 1.2.3.3 where some experiments are described in more detail.

1.2.3.2 Observables

When a subset of molecules out of an inhomogeneously broadened ensemble is selectively excited at low temperatures in its lowest energy electronic transition by a narrow band laser, the fluorescence spectrum will consist of narrow emission lines if the condition of weak electron-phonon coupling is fulfilled. Fluorescence line narrowing has been observed in a variety of dye-doped solids [40]. When the 0–0 transition is excited, the distance between the exciting laser frequency and the emission lines directly corresponds to the ground state vibrational frequencies. The width of the fluorescence lines is close to the width of the vibrational levels, determined by the dephasing time of the vibration and the inhomogeneous distribution of vibrational frequencies. The vibrational lines are a fingerprint of the emitting molecule and allow one to distinguish between molecules differing by chemical or isotopic substitution. Another important question concerns the sensitivity of the vibronic frequencies and intensities to the molecular environment. For polycyclic aromatic hydrocarbons – dyes that are used in SMS – the vibrational frequencies and vibronic intensities are only weakly sensitive to their environment [38]. Yet, as Myers et al. [38] pointed out, the pentacene molecule does seem to show some matrix-dependent vibronic intensity variations that may be potentially useful as a probe for environmentally induced skeletal distortions.

When we consider fluorescence spectra of single molecules, we first have to mention a fundamental difference from ensemble-averaged line-narrowed fluorescence spectra. Frequency selective measurements on large molecular populations are still affected by inhomogeneities because molecules which are accidentally degenerate with respect to the electronic transition frequency may still have different molecular environments and hence a spread of frequencies for a specific vibronic transition. By recording vibrationally resolved fluorescence spectra of an individual molecule, all inhomogeneities are removed and one hopes to get information about the molecule's local environment in the condensed phase as well as about the nature of the interactions between the molecule and its environment. Another motivation to do single molecule fluorescence spectroscopy is based on analytical reasons. The vibrational structure of a fluorescence spectrum can be employed to check the chemical identity of the single molecule investigated. In this sense, single molecule fluorescence spectroscopy would be an extremely sensitive method of detecting trace amounts of e.g. cancerogenic hydrocarbon contaminants [41].

1.2.3.3 Results of some specific systems

Pentacene in p-*terphenyl*

The first vibrationally resolved fluorescence spectra of individual molecules were reported for the "classic" mixed crystal system of pentacene in *p*-terphenyl by Tchenio, Myers and Moerner [42]. These spectra which were taken with an far from optimized setup demonstrated the feasibility of such measurements and several research directions that would be interesting to pursue were suggested by the authors. The same system was investigated later on by Fleury et al. [39] with an improved setup yielding spectra of higher quality. The experimental configuration used by the latter authors is similar to that shown in Fig. 8. Instead, however, of using an optical fiber to confine the exciting light, the authors employed a confocal scheme whereby a small region of the sample was illuminated by the same (high-quality) paraboloid that also served to collect the fluorescence light. The detection setup for the fluorescence light after the paraboloid was the same as in Fig. 8.

In Fig. 9 fluorescence spectra of two single pentacene molecules are displayed [39]. The experimental conditions used to record these spectra are given in the figure caption. We want to draw the attention of the reader to the high signal-to-noise ratio in these spectra which results from several factors. First, a precision manufactured paraboloid permitted efficient imaging of the fluorescence light on the entrance slit of the spectrograph. Second, single pentacene molecules in *p*-terphenyl are very stable with respect to their transition frequency, allowing long accumulation times which in this case reached up to 40 minutes. Third, the use of a liquid nitrogen-cooled CCD prevents build-up of excessive detector background. It is clear that spectra as shown in Fig. 9 can be easily used to identify the chemical and isotopic structure of the compound studied.

By comparing the single molecule fluorescence spectra with the spectra of an ensemble of pentacene molecules residing in the O_1 or O_2 site, respectively, molecule A in Fig. 9 was assigned to site O_1, and molecule B to site O_2. This assignment was possible for most molecules even in highly disordered crystals where the electronic transition frequency would give no information on the site relationship [39]. The main spectral lines were found to be remarkably constant from molecule to molecule which supports the above mentioned picture of the weak sensitivity of the vibrational frequencies and intensities of aromatic hydrocarbons to their environment. For some molecules intensity redistributions were observed which may relate to environment (defect) induced out-of-plane distortions of the pentacene skeleton. Another explanation considered by the authors invoked isotopic substitution by ^{13}C

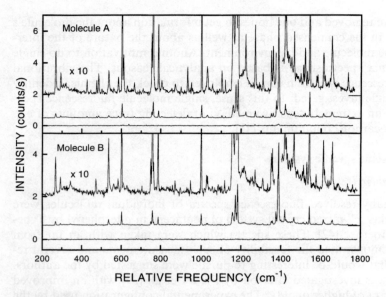

Figure 9. Fluorescence spectra of two single pentacene molecules in *p*-terphenyl recorded with a confocal setup (see text). Spectrum A was taken in 33 minutes with a sample contacted to a silica plate. The excitation frequency of this molecule coincided with the unperturbed O_1 site maximum. The spectrum presents all features of an O_1 molecule when compared to a bulk spectrum with only some small deviations [39]. Spectrum B was recorded over 40 minutes with a sample glued to a pinhole. The excitation frequency of the molecule was in the middle of the O_1–O_2 interval. According to its spectrum which shows some small differences to spectrum A it clearly belonged to site O_2 (from Ref. 39).

which also gives rise to vibrational frequency and intensity differences among molecules. Given the natural abundance of ^{13}C, the purely statistical probability that a pentacene molecule will contain at least one ^{13}C atom is after all 22%. Employing the QCFF/PI semiempirical method, intensity and frequency changes due to isotopic substitution of pentacene were predicted recently to be readily observable [42].

Terrylene in polyethylene

The most detailed investigation of single-molecule vibrational spectra has been carried out on the terrylene in polyethylene system by Tchénio et al. [38, 43, 44]. For this study the setup shown in Fig. 8 was used. Single molecule fluorescence spectra were obtained with excitation wavelengths both to the red and slightly to the blue of λ_{max} (569 nm) of the inhomogeneous distribution.

Since the chromophore terrylene is an almost unknown molecule spectroscopically, the interpretation of the fluorescence spectra had to rely entirely on comparison with the results of quantum chemical calculations (QCFF/PI + CISD [45]) that were performed by Myers [38] and colleagues. While the calculated frequencies were found to be uniformly too high, there was a fairly good correspondence between calculated and observed frequencies and intensities, particularly in the low

Figure 10. Vibrationally resolved fluorescence spectra of terrylene in polyethylene ($T = 1.4$ K). The "bulk" spectrum was obtained with excitation near the peak of the inhomogeneously broadened origin band. A–E represent spectra of different molecules which are described in the text. The average time to collect the single molecule spectra was several hundred seconds (from Ref. 38).

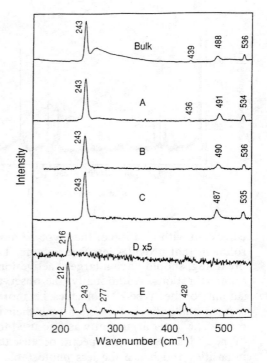

frequency region. Fig. 10 compares fluorescence spectra of single terrylene molecules and a bulk sample in the low frequency region. The main point to be discussed here is the interesting observation that two different types of single molecule fluorescence spectra were observed which were referred to as "type 1" and "type 2" spectra, respectively. Typical "type 1" spectra are represented by the spectra of molecules A, B and C in Fig. 10 and they closely resemble the "bulk-type" spectrum also shown in Fig. 10. About 10–20% of the molecules that were investigated in this particular sample showed a qualitatively different frequency and intensity pattern. These "type 2" molecules (D and E in Fig. 10) had their most intense line (212–216 cm^{-1}) shifted down in frequency by nearly 30 cm^{-1} relative to the "type 1" molecules where this intense low frequency mode appeared at 243 cm^{-1}. The overtone and combination bands involving this transition were shifted accordingly. Further prominent differences were as follows: A strong line found near 1562 cm^{-1} in the "type 1" spectra was shifted to slightly lower frequency, and there was only a single line near 1272 cm^{-1} where in "type 1" spectra a doublet appeared.

Based on the knowledge that polyethylene solidifies as a polycrystalline matrix in which small crystalline domains are connected by amorphous regions [46] and after ruling out other possibilities, the "type 1" spectra were tentatively assigned to molecules in or on the surface of the crystalline regions and "type 2" spectra to molecules in amorphous regions [43]. This assignment relied on a combination of vibrational spectroscopic and photophysical observations. The much higher frequency of the strong low-frequency mode in "type 1" spectra compared with "type 2" spectra was

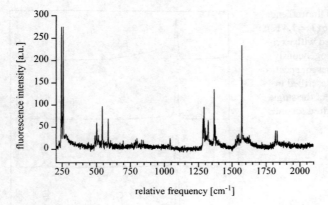

Figure 11. Fluorescence spectrum of a single terrylene molecule in *p*-terphenyl in site X_2 ($T = 1.4$ K). The accumulation time to record the complete spectrum was 600 seconds. The spectrum is in good agreement with a bulk spectrum of the X_2-site.

consistent with location of the "type 1" molecules in a more dense (crystalline) environment that provides less free volume. This vibration, involving overall long-axis breathing, should have a larger "activation volume" than most or all of terrylene's other vibrations. Additionally, the observation of apparently lower photophysical stability of the "type 2" molecules, i.e. more spectral diffusion and hole burning, also seemed qualitatively consistent with their residence in a more amorphous environment. The latter argument was also used to explain the almost complete absence of "type 2" features in bulk spectra because those were recorded with higher excitation intensities and hence the less photostable "type 2" molecules were thought to be more often subject to light-induced frequency changes. The assignment as discussed above was considered tentative and further corroborating evidence was required [38].

Terrylene in p-*terphenyl*

Fluorescence spectra of single terrylene molecules were also recorded recently in the crystalline matrix of *p*-terphenyl [29] using again a similar setup as shown in Fig. 8. Most of the observed vibrational frequencies coincide with those of terrylene in polyethylene with the exception of small frequency shifts. In Fig. 11 the fluorescence spectrum of a single terrylene molecule in site X_2 [29], which is the most photostable site, is displayed. The most important feature is the appearance of two lines of similar intensity at 244 and 256 cm^{-1}, respectively. This doublet has not to be confused with "type 1" and "type 2" spectra in terrylene/polyethylene as discussed in the preceding section because those stem from different molecules. In contrast, in the *p*-terphenyl crystal the doublet around 250 cm^{-1} is an inherent property of a single absorber, a fact which also can be deduced from bulk spectra through the appearance of the combination band at 500 cm^{-1}.

According to the theoretical study discussed above, the intense low frequency vibration around 250 cm^{-1} is attributed to a long axis stretch of the whole molecule. The authors of [38] also studied the influence of various possible static distortions of the terrylene molecule on its vibrational structure. A "boat" or "butterfly" distortion about the short axis such that the planes of the terminal naphthalenes make a dihedral angle of $\sim 20°$ redistributes the character of the intense low frequency mode, the energy of which was calculated to be 278 cm^{-1}, mainly into two modes at 266 and 292 cm^{-1}. Such a distortion which can explain the appearance of two lines in the low frequency region around 250 cm^{-1} (remember that the calculated frequencies came always out too high) may be induced by the crystal lattice as the undistorted terrylene molecules are assumed to be too large to substitutionally replace the *p*-terphenyl molecules in the crystal. A similar splitting of the prominent low frequency mode was also found in the other three sites of terrylene in *p*-terphenyl [29].

1.2.4 Quantum optical experiments

1.2.4.1 Introduction

Quantum optical experiments, by which we mean the investigation of the interaction between simple quantum systems and coherent light, have proven to be a successful tool to test basic concepts in quantum mechanics. Most of the experiments to date were performed with atoms or ions in vapors, beams or traps. An attractive feature of atoms relates to their relatively simple structure which is theoretically fairly well understood.

By storing a single ion in a quadrupole trap (Paul trap) and reducing its effective temperature by laser cooling an ideal situation for high precision optical spectroscopy is realized [47, 48], namely to investigate over long time periods the interaction of a monochromatic, coherent light beam with a single cold atom which is largely isolated from external pertubations, highly localized and not subject to Doppler broadening. Some important spectroscopic results with stored ions were the observation of quantum jumps [49–51], dark resonances with extremely small linewidths [52] and non-classical light fields [53].

Soon after the first recording of single molecule excitation spectra with good signal-to-noise ratio the question was posed as to which extent the same type of quantum optical experiments as were done with single ions stored in a RF trap could be performed with single molecules trapped in a solid. These considerations arose quite naturally as a single molecule clearly represents a single quantum system. The energy level structure of a typical aromatic hydrocarbon dye molecule (see Fig. 6), which typically consists of 30–40 atoms, at a first glance seems to be much more complicated than that of a two- or three-level atom. This is mainly due to the coupling of the intramolecular vibrations to the $S_0 \rightarrow S_1$ electronic transition investigated in SMS. The vibrational relaxation time of a large molecule in a solid, however, is extremely fast (picoseconds) and in the context of the experiments to be described below, the complicated level scheme of the molecule can be reduced to an effective three-level scheme for the much slower dynamics of the optical electron involving the singlet ground state (S_0), the first excited singlet state (S_1) and the first excited triplet state (T_1).

In optical experiments that can be also done on ensembles of molecules (e.g. external perturbations by electric fields [19, 20]) the importance of single molecule investigations results from the fact that a single absorber is exquisitely sensitive to its local environment and that any inhomogeneity has been removed. The quantum optical effects discussed below belong to another set of experiments as they do not have a counterpart in experiments on large ensembles. They would be either totally masked if many absorbers would contribute to the fluorescence signal or, as in the case of the dynamical Stark effect, are not accessible by ensemble line narrowing techniques in solids. Therefore, in the following we will talk about unique single absorber effects.

1.2.4.2 Quantum jumps

Quantum jumps in single trapped ions

The concept of quantum jumps which dates back to Bohr [54] describes the instantaneous transition of an electron between different energy levels of an excited atom. As the notion of quantum jumps appeared as a contradiction to a description in terms of the conventional formulation of quantum mechanics, there has been a lot of debate about the possibility of observing them. A famous attack against the concept of quantum jumps was initiated by Schrödinger in his provocative paper entitled "Are there quantum jumps?" [55].

In 1975 it was suggested by Dehmelt [56] that for a single ion in the so-called V configuration its quantum state could be monitored by looking at the level of the fluorescence signal. In this scheme an electric dipole allowed and a forbidden transition originating from a common ground state are driven resonantly by two laser fields. As long as the electron is driven in the allowed transition, strong fluorescence will be emitted. Eventually, with low probability, the electron will undergo a transition to the metastable state pumped by the second laser and will be shelved for a period of the order of its lifetime. During that time no photons will be detected. A single quantum jump back to the ground state will trigger the strong emission on the allowed transition. Therefore, a random sequence of bright and dark intervals will be observed. As a single quantum jump may trigger a large number ($10^8\,\mathrm{s}^{-1}$) of fluorescence events, this scheme was proposed to work as an atomic amplifier and, additionally, as an ultraprecise frequency standard because very long-lived forbidden transitions could be monitored. In 1986 three groups independently reported on the experimental observation of quantum jumps of single ions confined in radio frequency traps [49–51].

Dehmelt's original proposal not only gave rise to substantial experimental efforts to observe quantum jumps in single atomic ions but also occupied the minds of theoreticians trying to understand this fundamental process [57, 58]. Although a conventional quantum mechanical approach provides a complete understanding of the statistical properties of the system based on ensemble averages [59, 60], recent formulations as continuous quantum stochastic equations [61, 62] or the method of Monte Carlo wave functions [63, 64] deliver a more natural description of a system undergoing quantum jumps.

Quantum jumps of a single molecule

With reference to the energy level scheme in Fig. 6, we expect the distribution of the fluorescence photons emitted by a single molecule to be composed of bright intervals separated by dark periods. This distribution is also visualized in Fig. 6. As the photons have the tendency to be emitted in bunches this property is called photon bunching. It becomes immediately clear that after a dark period, when the electron is shelved in T_1, a single transition – a single quantum jump – from T_1 to S_0 will trigger the emission of a large number of fluorescence photons on the $S_0 \leftarrow S_1$-transition. Now the question arises whether it is possible to directly detect these quantum jumps between the electronic levels of a single molecule as random interruptions of the macroscopic fluorescence signal. The answer strongly depends on the internal photophysical dynamics of the system investigated and the experimental photon detection rate.

For a single terrylene molecule in p-terphenyl (site X_2 or X_4) approximate values of the ISC rates are $k_{23} \approx 10^3 \, \text{s}^{-1}$ and $k_{31} \approx 10^3 \, \text{s}^{-1}$ [6]. This means that we expect a molecule to undergo excitation-emission cycles in the singlet system for $\approx 2\,\text{ms}$ $(2/k_{23})$ and after ISC into the triplet state it will be shelved for $\approx 1\,\text{ms}$ $(1/k_{31})$. Considering the above parameters and a typical number of detected fluorescence photons of $\approx 500.000 \, \text{s}^{-1}$ for a single terrylene molecule at saturating intensities, any abrupt fluctuations of the fluorescence are easily observable when sampling the fluorescence intensity within time intervals of 100 μs [65]. By inspecting the same quantities for a single pentacene molecule in p-terphenyl (see [34, 36]) it is seen that such fluctuations in the fluorescence would not be observable using current detection schemes with efficiencies around 0.5%. In this case the fluorescence intensity autocorrelation function to be discussed in the next section is the technique of choice to unravel the dynamics of singlet–triplet quantum transitions in a single molecule.

In Fig. 12 the average fluorescence intensity of a single terrylene molecule in site X_4 of the p-terphenyl crystal is shown as a function of time [66]. To record such a trace, the molecule is excited in its absorption maximum with a fixed laser frequency and the fluorescence is detected with a multichannel scaler set to a sampling interval of 164 μs. In Fig. 12(b) a section of 1 s out of a trace with full length 26.8 s is shown. By magnifying a 80 ms region out of this trace the discrete interruptions of the fluorescence signal are clearly visible (Fig. 12(a)). The single molecule displays individual quantum jumps by turning on and of the strong fluorescence of the S_0–S_1 transition. This so-called random telegraph signal tells us the quantum state of the molecule i.e. whether it is in the singlet manifold or shelved in the triplet state.

Similar stochastic fluctuations of the fluorescence signal were also observed in other single molecule investigations, but they were caused by spontaneous or photo-induced rearrangements of the local environment of the molecule (spectral diffusion and spectral hole-burning, see Sections 1.4 and 1.5) which shift the single molecule absorption line in and out of resonance with the exciting laser frequency. These spectral jumps may not be confused with the singlet-triplet quantum jumps described here which are an intramolecular process.

In Fig. 13 we have plotted the distributions of interval lengths as derived from a trace with recording time of 26.8 s (see Fig. 12) for the situations when the molecule is in the singlet ("on" times) or in the triplet manifold ("off" times), respectively. These distributions allow the determination of the rates of the underlying processes [57, 65]. The distribution of "on" times was fitted by a single exponential while the data for the "off" times could be best approximated by a bi-exponential decay (see

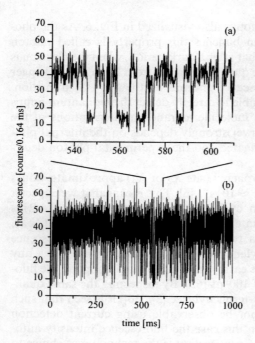

Figure 12. Fluorescence counts detected for a single terrylene molecule in *p*-terphenyl (site X_4) as a function of time. The quantum jumps are clearly visible as discrete interruptions of the fluorescence. The sample was cooled to 1.4 K and the molecule was excited with an (free space) intensity of 62.5 W/cm². (b), a 1 s section of a 26.8 s data set, (a), a region covering 80 ms displayed on a magnified scale.

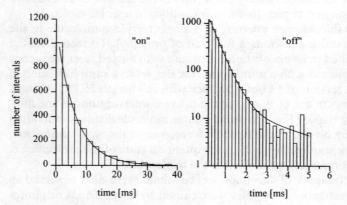

Figure 13. Distributions of the "on" and "off" intervals for a single terrylene molecule. The solid lines are an exponential ("on") and a biexponential ("off") fit to the data. Please note the logarithmic ordinate for the "off" intervals. The source of the data shown here was a trace of 26.8 s length, part of which is shown in Fig. 12.

also [67]). So far we had neglected the zero-field splitting of the triplet state which may give rise to a multi-level decay. From measurements with the correlation technique (see 1.2.4.3) it was expected that for terrylene the kinetics of two sublevels can be observed. The statistics for the "on" times, however, are not good enough to fit the data reliably with a bi-exponential though by appropriate plotting of the data one clearly sees deviations from the single exponential at long times. The fact that at most two out of the three sublevels are seen in the distributions is most probably caused by a behaviour typical for planar aromatic hydrocarbons. Usually, the two in-plane levels have similar rates which make them indistinguishable in purely optical experiments as described here. The out-of-plane level however, is typically characterized by slower rates leading to a bi-exponential distribution as shown in [6].

From the fits to the distributions we obtained $\tau_{on} = 6.3$ ms and $\tau_{off}^{(1)} = 0.43$ ms and $\tau_{off}^{(2)} = 3$ ms. The $\tau_{off}^{(i)}$ are independent on intensity and equal the lifetimes ($1/k_{31}$) of the corresponding sublevels whereas $\tau_{on} \sim 2/k_{23}$ only when the $S_0 \rightarrow S_1$ transition is fully saturated which was true for the data shown in Fig. 13. From a measurement of the fluorescence autocorrelation function (see next section) of the same molecule we obtained $k_{23} = 0.4 \times 10^3 \, s^{-1}$ and $k_{31}^{(1)} = 2.3 \times 10^3 \, s^{-1}$ and $k_{31}^{(2)} = 0.3 \times 10^3 \, s^{-1}$ which is in good agreement with the data derived from the direct quantum jump measurement.

The quantum dynamics of a single evolving quantum system is characterized by stochastic jumps between different electronic levels. For the case of a single terrylene molecule in a *p*-terphenyl matrix it could be shown that the quantum jumps between the singlet and triplet manifold are directly observable as discrete interruptions of the strong fluorescence signal. Such kind of text book experiments demonstrate the ability of SMS to reveal truly quantum-mechanical effects in large polyatomic molecules that were so far restricted to single atom investigations.

1.2.4.3 Fluorescence intensity autocorrelation function

Theoretical description of correlation effects in a single molecule

The time distribution of the fluorescence photons emitted by a single dye molecule reflects its intra- and intermolecular dynamics. One example are the quantum jumps just discussed which lead to stochastic fluctuations of the fluorescence emission caused by singlet–triplet quantum transitions. This effect, however, can only be observed directly in a simple fluorescence counting experiment when a system with suitable photophysical transition rates is available. By recording the fluorescence intensity autocorrelation function, i.e. by measuring the correlation between fluorescence photons at different instants of time, a more versatile and powerful technique is available which allows the determination of dynamical processes of a single molecule from nanoseconds up to hundreds of seconds. It is important to mention that any reliable measurement with this technique requires the dynamics of the system to be stationary for the recording time of the correlation function.

A thorough description of the correlation properties of classical and nonclassical light fields can be found in the excellent textbook by Loudon [68]. The correlation functions of the field $E(t)$ ($g^{(1)}(\tau)$: first order correlation function) and of the intensity $I(t)$ ($g^{(2)}(\tau)$: second order correlation function) of a light wave are defined classically:

$$g^{(1)}(\tau) = \langle E(t)E(t+\tau)\rangle / \langle E^2(t)\rangle$$

$$g^{(2)}(\tau) = \langle I(t)I(t+\tau)\rangle / \langle I(t)\rangle^2 \tag{10}$$

$$\langle I(t)I(t+\tau)\rangle = \lim_{T\to\infty} \frac{1}{T} \int_0^T I(t)I(t+\tau)dt$$

When we measure photons emitted by a single molecule using a quantum detector such as a photomultiplier tube, we interpret $I(t)$ to be the photon counting rate (fluorescence intensity at time t) and we can deduce the fluorescence intensity auto-correlation function $g^{(2)}(\tau)$ by counting the number of photon pairs separated by time τ over some integration time longer than all the characteristic times of the intensity fluctuations. In the following, the term "correlation function" refers to the properties of the normalized fluorescence intensity autocorrelation function. The given theoretical description of the correlation effects of a single molecule at low temperatures will closely follow the work of Orrit and colleagues [34, 69]. The probability of detecting a pair of photons in intervals $[t, t + \mathrm{d}t]$ and $[t + \tau, t + \tau + \mathrm{d}\tau]$ is proportional to the probability of occupation of the excited singlet state at time $t, \mathscr{P}(\mathscr{S}_1, t)$ and to the conditional probability that the molecule will be in this state at time $t + \tau, \mathscr{P}(\mathscr{S}_1, t + \tau | \mathscr{S}_0, t)$ [69]. In this context it is important to realize that the observation of the first photon projects the molecule into the ground state.

At this point we have arrived at the connection between the correlation function and the density matrix elements described in Section 1.2.2.3, because the conditional probability is just proportional to the matrix element $\sigma_{22}(\tau)$ corresponding to the transient solution of the optical Bloch equations for our model three-level system in Fig. 6. Then it follows,

$$g^{(2)}(\tau) = \frac{\mathscr{P}(\mathscr{S}_1, t)\mathscr{P}(\mathscr{S}_1, t + \tau | \mathscr{S}_0, t)}{\mathscr{P}(\mathscr{S}_1, t)^2} = \frac{\sigma_{22}(\tau)}{\sigma_{22}(\infty)}, \tag{11}$$

with,

$$\mathscr{P}(\mathscr{S}_1, t + \tau | \mathscr{S}_0, t) = \eta k_{21}^r \sigma_{22}(\tau) \tag{12}$$

In the above equations $\sigma_{22}(\infty)$ is the equilibrium population of the excited state S_1, η the overall detection yield and k_{21}^r the radiative singlet transition rate.

The expression for $\rho_{22}(\tau)$ can be obtained by solving the Bloch equations analytically by Laplace transform techniques [34]. The excited state population is obtained as a sum of complex exponentials:

$$\sigma_{22}(\tau) = \frac{\Omega^2}{2} \left\{ \frac{k_{31}\Gamma_2}{\lambda_1\lambda_2\lambda_3\lambda_4} + \sum_{i=1}^{4} \frac{(\lambda_i + k_{31})(\lambda_i + \Gamma_2)}{\lambda_i(\lambda_i - \lambda_j)(\lambda_i - \lambda_k)(\lambda_i - \lambda_l)} \exp(\lambda_i t) \right\} \tag{13}$$

The λ_i are the solutions of a fourth degree equation the coefficients of which are given explicitly in the literature [34, 70]. In the next paragraph we will present exact numerical simulations for $g^{(2)}(\tau)$ for a single three level system. Later we will use approximate analytical solutions for specific time regimes of $g^{(2)}(\tau)$ which render the physical interpretation of the observed effects more intelligible.

In Fig. 14 numerical simulations of the correlation function $g^{(2)}(\tau)$ for a single pentacene molecule are shown on a logarithmic time scale for different values of the Rabi frequency Ω [70]. In the simulation the photophysical parameters as determined for pentacene in *p*-terphenyl were used. At short times we recognize a corre-

Figure 14. Simulation of the fluorescence intensity autocorrelation function $g^{(2)}(\tau)$ for a single pentacene molecule for different values of the Rabi-frequency Ω: (a) 3 MHz; (b) 25.5 MHz; (c) 71.3 MHz (from Ref. 70).

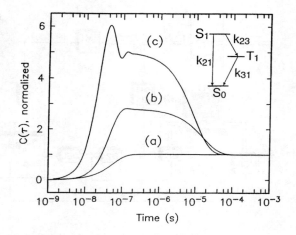

lation dip, i.e. $g^{(2)}(\tau)$ tends to zero for short separation times between photon pairs ($\tau \to 0$). For a classical coherent light field (see Eq. 10) the limiting value for $g^{(2)}(0)$ is unity [68]. A correlation function dropping below unity is indicative of a non-classical radiation field. The tendency of photons to be spaced out in time is phrased photon-antibunching. The phenomenon of antibunching is a true signature of a quantum field; however, for the case of a single molecule it can be nicely rationalized with the following simple picture. In the correlation measurement the conditional probability for the arrival of a photon at time t and the arrival of a second photon at time $t + \tau$ is recorded. After emission of a photon at time t the quantum system is prepared in its ground state since it just emitted a photon. The probability of emitting a second photon at t is zero because the molecule cannot emit from the ground state. On the average a time of half a Rabi period has to elapse to have a finite probability for the molecule to be in the excited state again and emit a second photon. Photon antibunching was first predicted by Carmichael and Walls [71] to be observable in the statistical properties of the resonance fluorescence of a single atom. Later on, this effect was actually observed experimentally in a weak beam of sodium atoms [72] and for a single stored Mg^+ ion [53].

It is seen in Fig. 14(a)–(c) that $g^{(2)}(\tau)$ increases from 0 ($\tau \to 0$) to higher values at longer times the actual value depending on the exciting intensity. At very long times ($\tau > 10^{-4}$ s) all traces reach the limiting value for uncorrelated events which is unity for the normalized correlation function. The steepness of the increase of $g^{(2)}(\tau)$ at short times clearly depends on the intensity because the excitation probability increases with increasing Ω. At the highest intensity the correlation function displays Rabi oscillations which represent the laser driven coherent motion between the singlet ground and excited state of the pseudospin. These oscillations are damped by energy and phase relaxation processes. The subsequent decay of the correlation function is caused by singlet-triplet transitions which induce the bunching of photons already mentioned where bursts of emissions are separated by dark intervals. At low exciting intensities (Fig. 14(a)) the contrast in the correlation function due to photon bunching vanishes. By decreasing the intensity, the average time spacing between

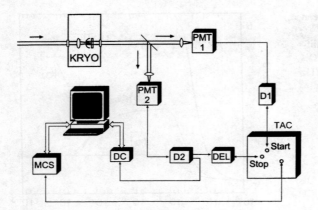

Figure 15. Experimental setup to measure the correlation function for a single molecule from nanoseconds up to hundreds of seconds. PMT: photomultiplier tube; D: discriminator; DEL: electronic delay; TAC: time-to-amplitude converter; MCS: multichannel scaler; DC: digital logarithmic correlator.

fluorescence photons comes close to the average length of dark periods which equals the triplet lifetime. In this situation the dark periods do not represent a characteristic feature in the correlation function and the contrast between bright and dark intervals goes to zero. Finally, it is an inherent property of the typical electronic three level structure of an organic dye molecule that its intensity correlation function displays the antibunching as well as the bunching phenomenon.

Experimental determination of the intensity correlation function

Before we discuss some specific examples of correlation measurements on single molecules we briefly want to describe in Fig. 15 an experimental setup to measure the correlation function over a wide time range. The photon stream emitted by a single molecule is first divided by a 50:50 beam splitter and the light focused onto the cathodes of two photomultiplier tubes. The amplified output pulses of the two phototubes are then normalized by two constant-fraction discriminators. The pulses from the discriminators activate the start and stop inputs of a time-to-amplitude converter (TAC) where the time delay between the two pulses is converted into a voltage. A fixed delay in the stop channel allows to measure the correlation function for "negative times". The voltage pulses from the TAC are finally stored in a multi-channel scaler. The reader may have realized that this part of the setup is basically equivalent to the famous Hanbury–Brown and Twiss intensity interferometer [73]. Using this equipment one can measure the distribution of consecutive photon pairs (delayed coincidences) between 1 nanosecond and 1 microsecond. In this time range and for low counting rates the distribution of consecutive photon pairs is indistinguishable from the full correlation function which describes the time delays between all photon pairs.

To measure accurately the (full) correlation function for longer times, the second output channel of one of the discriminators in Fig. 15 is used to simultaneously feed a digital logarithmic correlator. This device covers the time range from micro-seconds up to thousands of seconds whereby its minimum time resolution strongly depends on the after pulsing characteristics of the phototube. By employing the two

Figure 16. Normalized fluorescence intensity autocorrelation function $g^{(2)}(\tau)$ for a single terrylene molecule in *p*-terphenyl (site X_2) at 2 K measured over 9 orders of magnitude in time ($\bullet$). The drawn line is a simulation of the correlation function using appropriate photophysical parameters (from Ref. 27).

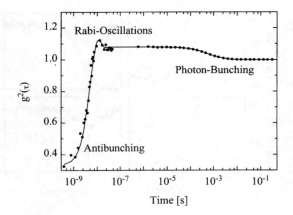

correlation schemes just discussed the dynamics of a single molecule can in principal be investigated over a time range covering eleven orders of magnitude.

Experimental results

Using the experimental setup (Fig. 15) described in the last paragraph, the fluorescence intensity correlation function was measured for a single terrylene molecule in *p*-terphenyl over nine orders of magnitude in time by the authors group [27]. The experimental trace in Fig. 16 clearly displays the characteristic features of photon antibunching and photon bunching. The onset of Rabi oscillations is clearly visible, too. We now want to discuss separately these three effects for different systems and what can be learned from such measurements.

Photon antibunching. Photon antibunching in SMS was first observed for single pentacene molecules in *p*-terphenyl [70]. The data for a single terrylene molecule in *p*-terphenyl shown in Fig. 16 clearly exhibit the anticorrelation at short times [27]. It is seen that $g^{(2)}(\tau)$ does not reach zero for $\tau \to 0$. This deviation from 0 is caused by accidental pair correlations of photons originating from laser light scattered at the host crystal. As the background increases approximately linearly in the spectrum and quadratically in the correlation function [34] with exciting intensity, the contrast of the antibunching is diminished with increasing laser intensity.

In the course of this paper we always implicitly took for granted that the experiments were performed with only a single molecule and not two or more. While there are several intelligible arguments which strongly support this assumption, the contrast of the antibunching signal gives unambiguous proof that we are really investigating an individual quantum system. The fluorescence excitation profiles of one and two pentacene molecules, the latter being so close in frequency space that their excitation lines overlap, are shown in Figs. 17(a) and (b), respectively. The corresponding experimental traces for $g^{(2)}(\tau)$ are given in Figs. 17(c) and (d). By adding a delay in the stop-channel (see Fig. 15), the correlation function could also be mea-

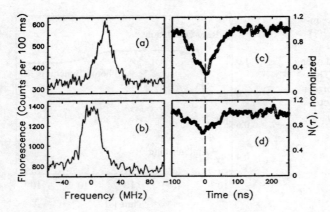

Figure 17. Fluorescence excitation spectra for one (a) and two (b) single pentacene molecules in p-terphenyl ($T = 1.4$ K). (c) and (d) show $g^{(2)}(\tau)$ for cases (a) and (b), respectively. Notice the difference of the contrast of the correlation functions in (c) and (d) (from Ref. 70).

sured at "negative" times which helps to establish time zero. It is seen that the contrast decreases by a factor of ≈ 2 when two molecules are excited simultaneously [70]. In general, the contrast of the antibunching signal for N independent emitters is proportional to $1/N$. It was pointed out already that any kind of correlation effect discussed here is strongest when only a single emitter is observed.

Rabi oscillations. The expression for the correlation function $g^{(2)}(\tau)$ at short times when the triplet state contribution is negligible can be written under certain approximations as follows:

$$g^{(2)}(\tau) = 1 - \exp\left(-\frac{(\Gamma_2 + k_{21})\tau}{2}\right)\left[\left(\frac{\Gamma_2 + k_{21}}{2\Omega}\right)\sin(\Omega\tau) + \cos(\Omega\tau)\right] \quad (14)$$

Besides the antibunching near $\tau = 0$, the above equation describes the Rabi oscillations at short times and contains via the dephasing rate Γ_2, $1/T_1$ and $1/T_2^*$ as damping terms. This equation demonstrates that under the valid assumption of constant T_1 – which is usually satisfied in the temperature range between 1 and 10 K – the pure dephasing time T_2^* can be extracted from the damping of the Rabi oscillations.

In Section 1.2.2.2 it was shown that the optical linewidth of a single terrylene molecule in p-terphenyl is broadened by librational mode induced pure optical dephasing (T_2^*-processes) when the temperature is increased. In Fig. 18, $g^{(2)}(\tau)$ is plotted at 2 K and 4 K for a single terrylene molecule in p-terphenyl [27]. It can be seen that at higher temperatures the Rabi oscillations become subject to much stronger damping because at elevated temperatures pure optical dephasing caused by coupling of the local mode to the electronic transition contributes to Γ_2. Good agreement was found between the values of T_2^* derived from the linewidth and the correlation measurement [27] which demonstrates that both techniques measure the same quantity. These results demonstrate how in single molecule experiments nano-

Figure 18. Fluorescence intensity autocorrelation function at short times for a single terrylene molecule in *p*-terphenyl (site X_2) at saturating intensities. The solid lines are fits of Eq. 14 to the data (◆): (a) 2 K; (b) 4 K. It is clearly seen that by raising the temperature the Rabi oscillations become subject to stronger damping which is caused by pure optical dephasing (from Ref. 27).

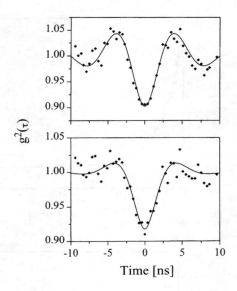

second time resolution can be achieved under continuous wave excitation by analyzing the statistical properties of the fluorescence photons.

Photon bunching. Quantum jumps between the singlet and triplet manifolds lead to photon bunching which appears as a decay of the correlation function at long times (see Figs. 14 and 16). Actually, this correlation effect was already demonstrated by Orrit and Bernard [2] in their first work on fluorescence detection of single molecules at low temperatures. When the triplet state is again represented by only one level, the decay of the normalized correlation function at long times when all coherences have vanished was shown to be a single exponential [34],

$$g^{(2)}(\tau) = 1 + C \exp(-\lambda t) \tag{15}$$

where the decay rate (λ) at resonance is related to the population (k_{23}) and depopulation (k_{31}) rates of the triplet state:

$$\lambda = k_{31} + k_{31} \frac{I}{I_s[(2k_{31}/k_{23}) + 1]} \left(1 + \frac{2k_{31}I}{I_s[2k_{31} + k_{23}]}\right) \tag{16}$$

The contrast C is given by:

$$C = (\lambda - k_{31})/k_{31} \tag{17}$$

By measuring the decay rate λ in dependence on the exciting intensity, the ISC parameters k_{23} and k_{31} can be obtained by fitting Eq. (16) to the data. At high

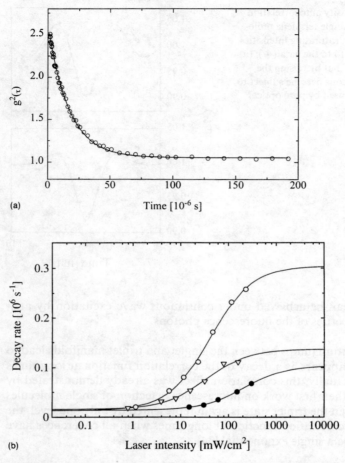

Figure 19. (a) Decay of the fluorescence intensity autocorrelation function $g^{(2)}(\tau)$ for a single penta-cene molecule in *p*-terphenyl at long times ($T = 1.4$ K). The solid line is a fit of Eq. 15 to the data. (b) Plot of λ versus logarithmic intensity for three different pentacene molecules. The solid lines are fits of Eq. 16 to the data. For high intensities ($\lambda \to k_{23}/2$) it can be seen that the population rate varies strongly from molecule to molecule (from Ref. 36).

intensity $(I \to \infty), \lambda \to k_{23}/2 + k_{31}$ and $C = k_{23}/2k_{31}$, while at low intensity $(I \to 0), \lambda \to k_{31}$ and $C \to 0$.

A typical decay curve of the correlation function in the µs time regime for a single pentacene molecule in *p*-terphenyl (site O_1) is shown in Fig. 19(a) on a linear time scale [36]. The decay rate can be well approximated by a single exponential (Eq. (15)). A plot of λ versus I is displayed in Fig. 19(b) for three different pentacene molecules. The drawn lines are fits of Eq. (16) to the experimental data. For penta-cene we have $k_{23} > k_{31}$ and at high intensity $\lambda \to k_{23}/2$. Thus, from Fig. 19(b) it can be seen that the population rate k_{23} varies strongly from molecule to molecule. These variations were first observed by Bernard et al. [34] and later on investigated in dependence on the disorder of the crystalline *p*-terphenyl host [36]. Following a model suggested by Kryschi et al. [74] it was concluded that the variations in the ISC parameter k_{23} reflect the local variations of defects which induce conformational distortions of the pentacene molecules. As ISC is forbidden by symmetry for perfectly planar pentacene molecules, such distortions are expected to change the ISC rates of the molecule. It is clear that the ability to measure distributions as described here is a mere result of SMS because from experiments on large populations one would only obtain the average value of the parent distribution which contains less information.

To keep Eqs. 15–17 more lucid, only one out of the three triplet sublevels was considered. Actually, in purely optical experiments with pentacene in *p*-terphenyl [36], the correlation decay seems to be dominated by a single level. A more detailed analysis of correlation experiments in connection with microwave induced magnetic resonance transitions between the triplet sublevels (see Section 1.6) allowed to determine the kinetics of all three triplet sublevels of a single pentacene molecule [69].

The decay of the correlation function for a single terrylene molecule in *p*-terphenyl due to photon bunching was already presented in Fig. 16. In this case a biexponential decay – not easily visible in Fig. 16 – is observed because two of the sublevels are sufficiently distinct with regard to their kinetics (see Section 1.2.4.2). The measurements and the analysis of the experimental data to determine the population and depopulation rates were done in analogy to the previous description for pentacene in *p*-terphenyl. The ISC rates were in fairly good agreement with those determined from quantum jump measurements discussed in Section 1.2.4.2. In contrast to pentacene in *p*-terphenyl, the terrylene molecules studied so far did not exhibit a large variation of the ISC rates between different molecules [6].

In general, the correlation technique as well as the quantum jump technique are powerful tools to unravel complicated molecular photophysical dynamics for a single absorber. This statement is exemplified by the investigation of the chromophore terrylene, for which no kinetical parameters of the triplet state were known from ensemble measurements. The ISC rates presented here were determined solely by experiments on single molecules. Actually, it would be quite difficult to measure absolute rates of photophysical ISC parameters by other techniques when the triplet quantum yield is smaller than 10^{-5}. Recently, fluorescence correlation spectroscopy was also proposed as an appropriate method for the determination of triplet parameters of fluorophores in solution [75]. Additionally, it is a helpful tool to investigate spectral diffusion of single absorbers as discussed in Sections 1.4 and 1.5.

1.2.4.4 Pump-probe experiments: light shift

In Section 1.3 the shift of the single molecule excitation line under the influence of a static electric field, the DC Stark effect, is discussed. The interaction of molecular electronic energy levels with a strong optical field is also expected to lead to level shifts and splittings and additionally to a change of relaxation rates. The shift of energy levels under optical excitation is called light shift or AC Stark effect where

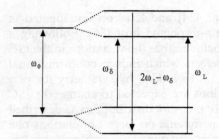

Figure 20. Energy-level diagram illustrating the interaction of a two-level atom with a strong near-resonant light field. ω_0 is the resonance frequency of the unshifted transition and ω_L is the laser frequency. The frequency of maximum probe absorption is shifted to ω_δ (light shift). The probe is amplified at $2\omega_L - \omega_\delta$ due to a three-photon process, involving absorption of two driving-field photons at ω_L and stimulated emission at $2\omega_L - \omega_\delta$ (from Ref. 79).

the quadratic effect of the pump laser field is amplified by near resonance with the transition.

The interaction of a strong light field with an atomic two-level system and the consequences for the emission and absorption spectrum of the atom were investigated theoretically by Mollow [76, 77]. To compute the absorption spectrum, Mollow evaluated in first order the weak-field induced dipole moment by means of Bloch equations in which the strong field was treated to all orders. Later on the theoretical predictions were verified experimentally [78, 79]. In case of a strong pump-field ($\Omega > \Gamma$) the emission spectrum consists of three components, the so-called Mollow triplet. The absorption spectrum recorded with a weak probe beam shows a more complex shape with absorption and amplification components under strong pump saturation. We want to consider the situation where the frequency of the pump beam is close to the atomic resonance frequency. In Fig. 20 the splitting of the energy levels by the pump laser field is displayed. The absorption spectrum is built up by three components: An absorbing transition at ω_δ, a three photon resonance at $2\omega_L - \omega_\delta$ and a third resonance at ω_L. The shift of the original resonance line from ω_0 to ω_δ under the influence of a near-resonant pump beam is called light shift. The multiphoton resonance at $2\omega_L - \omega_\delta$ does not lead to absorption but to an amplification of the probe beam. This nonlinear effect can be viewed as a transfer of photons from the pump to the probe beam. The line at ω_L corresponds to stimulated Rayleigh scattering.

An alternative description of the observed phenomena can be obtained by the "dressed atom" model [80]. In this approach the eigenvalues of the global system atom + field, the dressed states, are sought. The strong pump beam dresses the atom with laser photons which leads to energy levels split by the Rabi frequency. The resonances discussed in Fig. 20 then arise naturally as transitions between the (infinite) ladder of dressed states. The dressed atom model delivers a physically comprehensible description of atomic energy levels in strong fields and the reader is referred to Ref. 80 for a detailed presentation.

The strong inhomogeneous broadening of optical transitions in solids at low temperature prevented the observation of the AC Stark effect within an ensemble of molecules. In this case also line narrowing techniques did not provide a solution to realize the experiment. With the advent of single molecule spectroscopy the proper means to perform experiments of the pump-probe type on absorbers in solids without excessive interfering backgrounds were available. Due to the favourable photo-

Figure 21. Variations of the light shift with inverse of pump detuning for a single terrylene molecule in *p*-terphenyl. The pump intensities are as given in the figure. The two plots relate to two different molecules measured with different setups, under different illumination conditions (from Ref. 81).

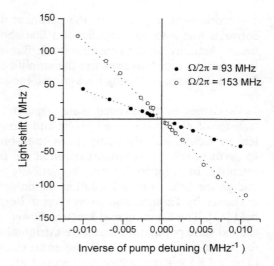

physical properties, single terrylene molecules in *p*-terphenyl (site X_2) were chosen as the most suitable system. The pump and probe beams were either provided by a single stabilized dye laser through frequency shifting with two acousto-optical modulators or by just employing two different dye lasers [81]. In a typical experiment the frequency of the strong pump laser is held fixed and the probe laser frequency is scanned over the molecular resonance line.

The light shift of the resonance line of a single terrylene molecule was measured with a near-resonant pump ($\Delta > \Gamma$) by either varying the pump intensity at a fixed detuning or vice versa. As can be derived from a perturbative approach as well as from the dressed atom model, the light shift δ for a near-resonant pump can be approximated ($\Delta > \Omega_1$) by the simple expression,

$$\delta = -\frac{\Omega_1^2}{2\Delta} \tag{18}$$

where Δ is the detuning between the pump frequency and the molecular resonance and Ω_1 the Rabi frequency of the pump beam. According to Eq. 18 the light shift δ should depend linearly on the pump intensity which could be verified experimentally [81]. In Fig. 21 the dependence of the light shift on the inverse pump detuning is shown for two molecules at two different pump intensities. The plots in Fig. 21 were also consistent with the simple formula in Eq. 18, at least within the range of intensities and detunings investigated by the authors [81]. From the slope of a straight line in Fig. 21 the Rabi frequency Ω_1 for the pump beam could be obtained. Ω_1 is related to the transition dipole moment μ and to its orientation. The ability to accurately determine Ω_1 from light shift data is an important accomplishment for the interpretation of optical saturation studies on single molecules, especially in the case of terrylene where μ is not known.

According to Fig. 20 and the pertinent discussion, besides the light-shifted absorption line also an amplification line should appear for a non-resonant pump beam. Actually, in the single molecule fluorescence excitation experiments population changes are measured and the amplification effect is expected to be very weak. Due to the large scattering background induced by the pump it was not observed so far.

Variations of the excited state population that gave indirect evidence for the transfer of photons between probe and pump beams were detected on single terrylene molecules with the pump beam on resonance [81]. The experimental conditions to perform these experiments required the probe and pump beam to be of similar intensity. In order to describe theoretically on-resonance data with arbitrary pump and probe fields, the full solution of the optical Bloch equations was numerically evaluated by Tamarat and co-workers in Bordeaux [81] following earlier work in this field [82]. The experimental spectra showed a W shape due to the variations of the excited state population that were attributable to the Autler–Townes splitting effect for weak probe power. All the measurements were done at low temperatures ($T = 1.4\,\mathrm{K}$) where dephasing processes are quenched. Simulations with additional dephasing gave spectral shapes that clearly deviated from those without dephasing [81]. It was therefore concluded to use such nonlinear pump–probe experiments to study optical dephasing and spectral diffusion for single molecules.

1.2.5 Conclusion and outlook

Optical excitation lines of single dye molecules in low temperature crystals that are not subject to spontaneous or light induced frequency jumps were shown to be ideal candidates for detailed spectroscopic investigations that require high excitation intensities or long accumulation times. We have described a variety of different experimental techniques which allow at the single molecule level to extract information that relates to host-guest interactions, intramolecular photophysical dynamics and structural properties of the dopant. While several of the experiments presented were mainly proof of principle, there is a wide range of interesting future applications some of them being given in the following paragraph.

Temperature dependent single molecule excitation spectra benefit from the fact that both the change in linewidth and line position can be extracted in a single experiment. Having access to the temperature dependent width and shift is expected to yield information on locally varying dynamic and static properties especially in disordered samples. Fluorescence spectra of single molecules allow to investigate the influence of ^{13}C isotopic substitution at virtually all positions of the chromophore without the need to synthesize specifically ^{13}C labeled compounds. Following the natural abundance of ^{13}C the purely statistical probability that e.g. a terrylene molecule contains at least one ^{13}C is $\sim 28\%$. It is clear that a meaningful assignment and interpretation of such spectra requires considerable theoretical efforts. Quantum optical measurements on single molecules are interesting on the one side because they deliver a clear picture of fundamental aspects of light–matter interactions. On the other side they have been proven to reflect also the dynamics of the local environment of the absorbers. The long term importance of quantum optical effects in the field of SMS in solids may largely depend on the extent to which they are sensitive to interactions between the dopant and the host.

Finally, taking into account all the additional methods that are available now (and that are presented in the other chapters of this book) it is a fascinating perspective to see them combined in the investigation of one single absorber. Such measurements will allow insights with unprecedented sensitivity into the local host–guest interactions in doped solids. While such investigations are difficult and time consuming, they are definitely in the realm of what is possible. In the authors group it was shown that one and the same single molecule (in a very low concentration crystal at low temperatures) could be studied over several weeks without loosing its identity.

Acknowledgements

The research of T. B., S. K. and C. B. that was described in this Section has been supported by the Deutsche Forschungsgemeinschaft. We are indebted to our colleagues R. Kettner, J. Tittel, S. Mais and F. Kulzer for their valuable contributions. T. B. acknowledges the very fruitful collaboration with W. E. Moerner at IBM.

References

[1] W. E. Moerner and L. Kador, *Phys. Rev. Lett.* **62**, 2535, (1989).
[2] M. Orrit and J. Bernard, *Phys. Rev. Lett.* **65**, 2716, (1990).
[3] T. E. Orlowski and A. H. Zewail, *J. Chem. Phys.* **70**, 1390, (1979).
[4] D. A. Wiersma in *Adv. Chem. Phys.*, Vol. 47 (Eds.: J. Jortner, R. D. Levine, and S. A. Rice), Wiley, New York, 1981.
[5] J. H. Meyling and D. A. Wiersma *Chem. Phys. Lett.* **20**, 383, (1973).
[6] S. Kummer, Th. Basché, and C Bräuchle, *Chem. Phys. Lett.* **229**, 309, (1994); *Chem. Phys. Lett.* **232**, 414, (1995).
[7] See e.g.: *Spectroscopy and Excitation Dynamics of Condensed Molecular Systems* (Eds.: V. M. Agranovich and R. M. Hochstrasser), North-Holland, Amsterdam, 1983.
[8] J. L. Skinner, *Ann. Rev. Phys. Chem.* **39**, 463, (1988).
[9] K. K. Rebane, *Impurity Spectra of Solids*, Plenum Press, New York, 1970.
[10] W. P. Ambrose, Th. Basché, and W. E. Moerner, *J. Chem. Phys.* **95**, 7150, (1991).
[11] W. E. Moerner, T. Plakhotnik, T. Irngartinger, M. Croci, V. Palm, and U. P. Wild, *J. Phys. Chem.* **98**, 7382, (1994).
[12] R. Kubo in *Fluctuation, Relaxation and Resonance in Magnetic Systems*, (Ed.: D. Ter Harr), Oliver and Boyd, Edinburgh, 1965.
[13] Th. Basché, W. P. Ambrose, and W. E. Moerner, *J. Opt. Soc. Am. B* **9**, 829, (1992).
[14] L. Fleury, A. Zumbusch, M. Orrit, R. Brown, and J. Bernard, *J. Luminescence* **56**, 15, (1993).
[15] J. Tittel, R. Kettner, Th. Basché, C. Bräuchle, H. Quante, and K. Müllen, *J. Luminescence*, **64**, 1, (1995).
[16] A. M. Stoneham, *Rev. Mod. Phys.* **41**, 82, (1969).
[17] L. Kador, *J. Chem. Phys.* **95**, 5574, (1991).
[18] D. L. Orth, R. J. Mashl, and J. L. Skinner, *J. Phys.-Cond. Matter* **5**, 2533, (1993).
[19] U. P. Wild, F. Güttler, M. Pirotta, and A. Renn, *Chem. Phys. Lett.* **193**, 451, (1992).
[20] M. Orrit, J. Bernard. A. Zumbusch, and R. I. Personov, *Chem. Phys. Lett.* **196**, 595, (1992).

[21] M. Croci, H. J. Müschenborn, F. Güttler, A. Renn, and U. P. Wild, *Chem. Phys. Lett.* **212**, 71, (1993).

[22] L. R. Narasimhan, K. A. Littau, D. W. Pack, Y. S. Bai, A. Elschner, and M. D. Fayer, *Chem Rev.* **90**, 439, (1990).

[23] See e.g.: *Persistent Spectral Hole-Burning: Science and Applications*, (Ed.: W. E. Moerner), Springer, Berlin, 1988.

[24] See e.g.: *Laser Spectroscopy of Solids*, Top. Appl. Phys. Vol. 49, (Eds.: W. M. Yen and P. M. Selzer), Springer, New York, 1986.

[25] F. P. Burke and G. J. Small, *Chem. Phys.* **5**, 198, (1974).

[26] W. H. Hesseelink and D. A. Wiersma, *J. Chem. Phys.* **73**, 648, (1980).

[27] S. Kummer, S. Mais, and Th. Basché, *J. Phys. Chem.*, **99**, 17078, (1995).

[28] F. G. Patterson, W. L. Wilson, H. W. H. Lee, and M. D. Fayer, *Chem. Phys. Lett.* **110**, 7, (1984).

[29] S. Kummer, PhD thesis, University of Munich, 1996.

[30] C. B. Harris, *J. Chem. Phys.* **67**, 5607, (1977).

[31] S. Völker, R. M. Macfarlane, and J. H. van der Waals, *Chem. Phys. Lett.* **53**, 8, (1978).

[32] J. Birks, *Photophysics of Aromatic Molecules*, Wiley-Interscience, London, 1970.

[33] S. P. McGlynn, T. Azumi, and M. Kinoshita, *Molecular Spectroscopy of the Triplet State*, Prentice Hall, Englewood Cliffs, 1969.

[34] J. Bernard, L. Fleury, H. Talon, and M. Orrit, *J. Chem. Phys.* **98**, 850, (1993).

[35] H. de Vries and D. A. Wiersma, *J. Chem. Phys.* **72**, 1851, (1980).

[36] Th. Basché, S. Kummer, and C. Bräuchle, *Chem. Phys. Lett.* **225**, 116, (1994).

[37] T. Plakhotnik, W. E. Moerner, V. Palm, and U. P. Wild, *Optics Comm.* **114**, 83, (1994).

[38] A. B. Myers, P. Tchenio, M. Z. Zgierski, and W. E. Moerner, *J. Phys. Chem.* **98**, 10337, (1994).

[39] L. Fleury, Ph. Tamarat, B. Lounis, J. Bernard, and M. Orrit, *Chem. Phys. Lett.* **236**, 87, (1995).

[40] R. I. Personov in *Spectroscopy and Excitation Dynamics of Condensed Molecular Systems* (Eds.: V. M. Agranovich and R. M. Hochstrasser), North-Holland, Amsterdam, 1983.

[41] R. Jankowiak and G. J. Small, *Anal. Chem.* **61**, 1023, (1988).

[42] P. Tchenio, A. B. Myers, and W. E. Moerner, *J. Phys. Chem.* **97**, 2491, (1993).

[43] P. Tchenio, A. B. Myers, and W. E. Moerner, *Chem. Phys. Lett.* **213**, 325, (1993).

[44] A. B. Myers, P. Tchenio, and W. E. Moerner, *J. Lumin.* **58**, 161, (1994).

[45] F. Zerbetto, M. Z. Zgierski, F. Negri, and G. Orlandi, *J. Chem. Phys.* **89**, 3681, (1988).

[46] P. J. Flory, *Pure Appl. Chem.* **56**, 305, (1984).

[47] H. Dehmelt in *Advances in Laser Spectroscopy* (Ed.: F. T. Arecchi, F. Strumia, and H. Walther), Plenum, New York, 1983.

[48] W. M. Itano, J. C. Bergquist, and D. J. Wineland, *Science* **237**, 612, (1987).

[49] W. Nagourney, J. Sandberg, and H. Dehmelt, *Phys. Rev. Lett.* **56**, 2797, (1986).

[50] Th. Sauter, W. Neuhauser, R. Blatt, and P. E. Toschek, *Phys. Rev. Lett.* **57**, 1696, (1986).

[51] J. C. Bergquist, R. G. Hulet, W. M. Itano, and D. J. Wineland, *Phys. Rev. Lett.* **57**, 1699, (1986).

[52] I. Siemers, M. Schubert, R. Blatt, W. Neuhauser, and P. E. Toschek, *Europhys. Lett.* **18**, 139, (1992).

[53] F. Diedrich and H. Walther, *Phys. Rev. Lett.* **58**, 203, (1987).

[54] N. Bohr, *Phil. Mag.* **26**, 1, (1913).

[55] E. Schrödinger, *Brit. J. Phil Sci.* **3**, 239, (1952).

[56] H. G. Dehmelt, *Bull. Am. Phys. Soc.* **20**, 60, (1975).

[57] R. J. Cook and H. J. Kimble, *Phys. Rev. Lett.* **54**, 1023, (1985).

[58] J. Javanainen, *Phys. Rev. A* **33**, 2121, (1986).

[59] A. Schenzle and R. G. Brewer, *Phys. Rev. A* **34**, 3127, (1986).

[60] D. T. Pegg and P. L. Knight, *Phys. Rev. A* **37**, 4303, (1988).

[61] N. Gisin and I. C. Percival, *J. Phys. A* **25**, 5677, (1992).

[62] C. W. Gardiner, A. S. Parkins, and P. Zoller, *Phys. Rev. A* **46**, 4363, (1992).

[63] K. Molmer, Y. Castin, and J. Dalibard, *J. Opt. Soc. Am. B* **10**, 524, (1993).

[64] B. M. Garraway, M. S. Kim, and P. L. Knight, *Opt. Commun.* **117**, 560, (1995).

[65] Th. Basché, S. Kummer, and C. Bräuchle, *Nature* **373**, 132, (1995).

[66] F. Kulzer, *Diploma thesis*, University of Munich, 1995.

[67] M. Vogel, A. Gruber, J. Wrachtrup, and C. von Borczyskowski, *J. Phys. Chem.* **99**, 14915, (1995).

[68] R. Loudon, *The Quantum Theory of Light*, Oxford University Press, Oxford, 1983.

[69] R. Brown, J. Wrachtrup, M. Orrit, J. Bernard, and C. von Borczyskowski, *J. Chem. Phys.* **100**, 7182, (1994).

[70] Th. Basché, W. E. Moerner, M. Orrit, and H. Talon, *Phys. Rev. Lett.* **69**, 1516, (1992).

[71] H. J. Carmichael and D. F. Walls, *J. Phys. B* **9**, L43, (1976).

[72] H. J. Kimble, M. Dagenais, and L. Mandel, *Phys. Rev. Lett.* **39**, 691, (1977).

[73] R. Hanbury Brown and R. Q. Twiss, *Proc. Roy. Soc. A* **242**, 300, (1957); *Proc. Roy. Soc. A* **243**, 291, (1958).

[74] C. Kryschi, H. C. Fleischmann, and B. Wagner, *Chem. Phys.* **161**, 485, (1992).

[75] J. Widengren, Ü. Mets, and R. Rigler, *J. Phys. Chem.* **99**, 13368, (1995).

[76] B. R. Mollow, *Phys. Rev.* **188**, 1969, (1969).

[77] B. R. Mollow, *Phys. Rev. A* **5**, 2217, (1972).

[78] R. E. Grove, F. Y. Wu, and S. Ezekiel, *Phys. Rev. A* **15**, 227, (1977).

[79] F. Y. Wu, S. Ezekiel, M. Ducloy, and B. R. Mollow, *Phys. Rev. Lett.* **38**, 1077, (1977).

[80] C. Cohen-Tannoudji, J. Dupont-Roc, and G. Grynberg, *Atom–Photon Interactions*, Wiley, New York, 1992.

[81] Ph. Tamarat, B. Lounis, J. Bernard, M. Orrit, S. Kummer, R. Kettner, S. Mais, and Th. Basché, *Phys. Rev. Lett.* **75**, 1514, (1995).

[82] G. S. Agarwal and N. Nayak, *J. Opt. Soc. Am. B* **1**, 164, (1984).

1.3 Polarization and Lifetime Measurements, External Perturbations and Microscopy

M. Croci, H.-J. Müschenborn, U. P. Wild

1.3.1 Introduction

In this chapter we continue the discussion of experiments performed on stable single molecules and introduce the technique of single molecule fluorescence microscopy which will be applied to the study of spectral dynamics.

Linearly polarized light represents a powerful tool to investigate the orientation of molecules. The exact orientation of a single molecule can be determined even in amorphous samples, since no averaging over a distribution of orientations takes place. In the second section single molecule polarization spectroscopy is demonstrated on the "classical" system pentacene in *p*-terphenyl and is applied to the study of the correlation between the observed pentacene spectral sites and the crystallographic substitutional sites. In Section 1.3.3, the measurement of the time-resolved fluorescence decay curve of single pentacene molecules doped in *p*-terphenyl, following repetitive excitation with laser pulses, is presented. The high sensitivity of zero phonon lines to the local environment and to very small perturbations allows one to study the effect of static external electric fields as described in Section 1.3.4. Both linear and quadratic Stark shifts were observed. The near-field technique was also applied to low-temperature single-molecule spectroscopy. Section 1.3.5 describes the application of pressure causing a frequency shift of single-molecule resonance lines. Finally, we will present (in Section 1.3.6) the development of single-molecule microscopy using different setups for the imaging optics. Besides the localization of the emitting molecule, microscopy will be applied as a powerful parallel technique to study many molecules at the same time, still avoiding ensemble averaging by separating them spatially.

1.3.2 Spectroscopy with polarized light

1.3.2.1 Introduction

In the gas phase or in isotropic hosts, such as liquids, polymers or glasses, the transition dipole moments of the guest molecules exhibit random orientations. In anisotropic systems, like crystals and stretched polymer films, there is usually a correlation between the orientation of the molecule and the geometry of the host matrix. The probability of electric dipole absorption depends on the square of the scalar product of the transition dipole moment vector with the electric vector of the light. Spectroscopy with polarized light can be used to determine the orientation of the molecular transition dipole moment and from this information can be obtained about the embedding geometry.

In the following we describe how polarized light can be applied to single-molecule spectroscopy and show results on the system pentacene in *p*-terphenyl. The 24 possibilities for an assignment of the four observed spectroscopic pentacene sites (named O_1, O_2, O_3, and O_4) to the crystallographic substitutional positions in the *p*-terphenyl unit cell (named M_1, M_2, M_3, and M_4) have been reduced to only 2 possibilities. The birefringence of the host crystal allows one to determine the depth of the emitting chromophore, relative to the crystal surface [1–3]. In combination with fluorescence microscopy (see last section) this method allows a three-dimensional localization of an emitting molecule.

The *p*-terphenyl molecule contains three conjugated phenyl rings connected in a linear (*para*) arrangement. From simple considerations one can see that there are two opposing contributions which determine the geometry of this molecule: the steric repulsion of the *ortho* hydrogens drives the central phenyl ring out of the plane determined by the two outer rings; on the other hand the stabilization of the π electron system tends to favor a planar structure. At high temperatures (in the crystalline phase) the central phenyl ring vibration occurs in a symmetric double well potential, the potential barrier height being about $4.6\,\mathrm{kJ\,mol^{-1}}$ [4]. The average geometry at room temperature is therefore planar and the unit cell is monoclinic with two molecules per unit cell (axes *a*, *b*, *c*). The two different orientations of the molecule in the unit cell are characterized by the angle Θ between the plane of the molecule determined by the outer rings (M-axis) and the *b*-axis, its value is $32.7°$ or $-32.7°$. The upper drawing in Fig. 1 shows the projection of the unit cell onto the *a/b*-plane for the high-temperature monoclinic unit cell.

At 193 K there is a phase transition to a triclinic unit cell. This disorder–order transition results from the localization of the central ring of each molecule in one of the two minima of the double well potential. Below 193 K the potential becomes asymmetric due to steric repulsion. For better comparison, however, a pseudo-monoclinic unit cell with $a' \sim 2a$, $b' \sim 2b$, $c \sim c$ and 8 molecules will be considered (following the nomenclature of Baudour et al. [5]). There are four different inequivalent positions M_1, M_2, M_3, and M_4 characterized by a different angle of the central ring of the *p*-terphenyl molecule relative to the plane formed by the two outer rings, as shown in the lower drawing of Fig. 1. Here, the orientation of the two outer rings of *p*-terphenyl is drawn as a thick bar (M-axis), whereas the position of the central ring (M'-axis) is given by the thin bar. For the interpretation of x-ray diffraction data at low temperature Baudour et al. [5] have postulated the existence of two equally probable domains α and β related by a symmetry plane: molecules in the same sites but in different domains have opposite orientation as can be seen in Fig. 1. The data about the orientation of the inequivalent positions are summarized in Table 1 for the room and low temperature phases. The differences are in the orientation of the coplanar outer rings (M-axis) relative to the *b'*-axis and in the torsion angle of the central ring (equivalent to the orientation of the M'-axis relative to the *b'*-axis).

Figure 1. The *p*-terphenyl crystal structure at room temperature and below the phase transition at 193 K. Shown is the projection of the short axes M of the molecules onto the *a/b*-plane (respectively *a'/b'* for the low temperature phase).

At room temperature (upper drawing) the average structure is planar. The orientation is given by the angle Θ between the molecular plane and the *b*-axis. There are two possibilities: $\Theta = 32.7°$ and $\Theta = -32.7°$.

The central rings stabilize, in the low-temperature phase, in one of the potential wells, whereas the outer rings move slightly in the opposite direction. The orientation of the two outer rings of *p*-terphenyl is drawn as a thick bar (M-axis), whereas the position of the central ring (M'-axis) is given by the thin bar. There are now four inequivalent positions M_1, M_2, M_3, and M_4. For the interpretation of X-ray diffraction data at low temperature Baudour et al. [5] have postulated the existence of two equally probable domains α and β related by a symmetry plane. Figure adapted from Ref. 2.

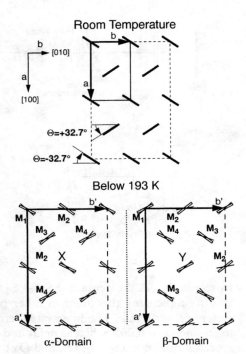

Table 1. Orientation of the short axes (M and M') of the phenyl rings of *p*-terphenyl relative to [010]-direction (*b* respectively *b'*-axis). From [2].

Room Temperature	Below Phase Transition (193 K) α-Domain			
M-axis	Position	M-axis outer rings	Central ring (M'-Axis)	Torsional angle of the central ring
-32.7°	M_1	-27.2°	-42.3°	15.1°
-32.7°	M_2	-36.5°	-10.2°	26.3°
+32.7°	M_3	+37.5°	+19.5°	18.0°
+32.7°	M_4	+27.0°	+50.3°	23.3°

As already mentioned in previous chapters, pentacene molecules substitute for *p*-terphenyl molecules. We assume that the pentacene molecule has the same orientation as the removed *p*-terphenyl (the short axis of the pentacene is aligned with the M-axis of the substituted *p*-terphenyl and their long axes are aligned) and that it is not strongly deformed. The transition dipole moment of the pentacene molecule introduced is oriented along its short axis. The different substitutional positions (M_1, M_2, M_3 or M_4) of the crystal host cause a different solvent shift, for the four spec-

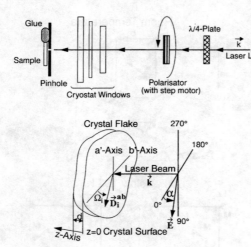

Figure 2. The linear polarized laser light is converted to circularly polarized light by a $\lambda/4$ wave plate and then is linearly repolarized in the desired direction by a computer controlled polarizer. The excitation beam passes through a pinhole and reaches the sample. The windows of the cryostat had to be carefully checked at the experimental temperature for depolarization properties. The lower part clarifies the definitions and orientations of the properties discussed in the text. Figure adapted from Ref. 1.

troscopic sites of the pentacene $S_1 \leftarrow S_0$ transition, named O_1, O_2, O_3, or O_4. These spectroscopic sites also differ in other photophysical properties, such as the intersystem crossing rate. The intersystem crossing quantum yield of pentacene at 1.4 K in O_1 and O_2 is 0.005, whereas in O_3 and O_4 it is about 0.6 [6]. The assignment of the spectroscopic sites O_1, O_2, O_3, and O_4 to the corresponding substitutional sites M_1, M_2, M_3, and M_4 will be discussed in the following sections.

1.3.2.2 Experimental

The sample, consisting of a thin flake, was carefully glued to a stainless steel plate with a 5 µm pinhole in its center. The excitation volume was smaller than 100 µm³. The plate was mounted onto a holder and immersed in liquid helium. The setup contains no optical components in the liquid helium path which affect the polarization state of the light. Fig. 2 shows how the polarization plane of the excitation light is controlled. Circularly polarized light is produced from the incoming linearly polarized laser light using a $\lambda/4$-wave-plate. A second polarizer, mounted on a computer controlled step-motor, polarizes the light again linearly and allows one to turn the angle of the polarization plane. The light passing through the pinhole and the cryostat windows showed no depolarization even at liquid helium temperature.

The p-terphenyl flake is mounted with its a'/b'-plane parallel to the pinhole plate. The linear polarized exciting light travels perpendicular to the a'/b'-plane, propagating along the z-axis of the coordinate system. The angle between the b'-axis and the electric field vector $\vec{E}$ before the light enters the sample is called α and gives the orientation of the polarization plane relative to the b'-axis. The projection of the transition dipole moment $\vec{D}_i$ of molecule i onto the a'/b'-plane is named $\vec{D}_i^{ab}$. The orientation of the transition dipole moment is defined as the angle Ω_i between $\vec{D}_i^{ab}$ and the b'-axis (lower part of Fig. 2).

In the beginning of a typical experiment the polarization plane of the light is parallel to the b'-axis. While recording the fluorescence, the polarization plane is rotated

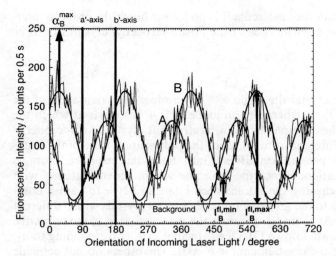

Figure 3. Fluorescence intensity of two different molecules as a function of the polarization of the exciting light. The molecules are located near the center of site O_1. The fluorescence intensity shows a sinusoidal modulation. For molecule B the maximum of the fluorescence intensity α_B^{max} is shifted with respect to molecule A. There is no position were the fluorescence of molecule B vanishes completely. The fluorescence curves are fitted to a sinusoidal function. The fit parameters yield the values for $I_i^{fl,max}$, $I_i^{fl,min}$, and α_i^{max}. Figure adapted from Ref. 1.

with the stepmotor-controlled retardation plate. This procedure allows the determination of the amplitude and the phase of the modulated fluorescence intensity (Fig. 3).

1.3.2.3 Results and discussion

Modulation of the fluorescence intensity

Fig. 3 illustrates the result of the polarization experiment for two molecules spectrally located in the O_1 region around 592.32 nm. Plotted is the angle α between the b'-axis of the crystal and the electric field vector of the linearly polarized excitation light, before entering the sample, against the fluorescence intensity. Both molecules show a sinusoidally modulated fluorescence intensity, but have different phases and amplitudes.

The phase α_i^{max} corresponds to the angle of the incident polarization at which the fluorescence intensity reaches its maximum value. The background signal was checked by measuring the signal a few MHz out of resonance and was found to be equal for molecules A and B. Its magnitude depends on the dark count rate of the photomultiplier and the scattered light. Molecule A has the maximal fluorescence at an angle of the incident polarization $\alpha_A^{max} = -24° \pm 7°$. Furthermore, if the polarization is turned 90° away from α_A^{max}, molecule A stops fluorescing. The fluorescence maximum I_B^{max} of molecule B is larger than the one of A and there is no angle at which its fluorescence completely vanishes. This suggests that the laser polarization at the location of molecule B is no longer linear. In order to inter-pret these data correctly the birefringence of the host *p*-terphenyl crystal has to be considered.

Theoretical interpretation

For a non-birefringent host crystal the angle α_i^{max} determines the orientation of the transition dipole moment Ω_i. Furthermore an angle $\alpha_i^{max} + 90°$ exists at which the fluorescence intensity vanishes completely. In a birefringent crystal like *p*-terphenyl the angle α_i^{max} is generally different from Ω_i, since the polarization state of the excit-ing light depends also on the penetration distance into the crystal, measured from its surface to the molecule location. Hence the molecule will in general interact with elliptically polarized light, causing the fluorescence intensity never to vanish com-pletely. *p*-Terphenyl belongs to the group of optically biaxial crystals with 3 different principal refractive indices. At room temperature the principal axes of the refractive index ellipsoid $n_1 < n_2 < n_3$ are known [7]. One principal axis, corresponding to n_2, has the same orientation as the crystallogaphic b'-axis. The others do not coincide with the crystallographic axes. Simple models [1, 2, 8] show that the fluorescence intensity I_i^{fl} is modulated sinusoidally. The modulation depth m_i determines the maximum and minimum fluorescence intensity, and strongly depends on the loca-tion z_i of the molecule inside the crystal:

$$m_i = \frac{I_{max} - I_{min}}{I_{max} + I_{min}} \tag{1}$$

The maximum modulation depth, $m_i = 1$, is observed for molecules (such as mol-ecule A) whose fluorescence intensity vanishes completely for certain excitation angles α. A fit to the experimental data yields the phase angle α_i^{max} and the maximum and minimum fluorescence intensities. From these, the modulation depth m_i can be calculated according to Eq. (1). The orientation of the molecular dipole moment Ω_i is then given by:

$$\begin{aligned} \Omega_i &= +\tfrac{1}{2} \times \arccos[m_i \times \cos 2\alpha_i^{max}], \qquad \text{for } \alpha_i^{max} > 0 \\ \Omega_i &= -\tfrac{1}{2} \times \arccos[m_i \times \cos 2\alpha_i^{max}], \qquad \text{for } \alpha_i^{max} < 0 \end{aligned} \tag{2}$$

The meaning of Eq. (2) is visualized and summarized in Fig. 4. The orientation α_i^{max} of the polarization plane before the sample at the maximum fluorescence $I_i^{fl, max}$, which can be directly obtained from the experiments, is plotted versus the location z_i of molecule i (relative to the crystal surface) for different orientations of the tran-sition dipole moment Ω_i (in steps of 10°). For a chromophore close to the surface $z_i \approx 0$ or located around $z_i \approx 8.8\,\mu m$, the measured α_i^{max} corresponds to the ori-entation of the transition dipole moment Ω_i. The birefringence of the crystal regen-erates the initial direction of the light polarization after 8.8 μm.

Let us consider the black trace for a molecule with $\Omega_i = +30°$, positions ① and ⑥ corresponding to the case $\alpha_i^{max} = \Omega_i$ discussed above. Moving deeper into the crystal (position ②) α_i^{max} gets smaller than Ω_i, until, at $z_i \cdot \Delta = \pi/2$ (position ③), the max-

Figure 4. Plot of α_i^{max}, the orientation of the polarization plane, before the sample, at the maximum fluorescence $I_i^{fl,max}$, which is directly obtained during the experiments, versus the location z_i of molecule i (relative to the crystal surface) for different orientations of the transition dipole moment Ω_i (in steps of 10°). For $z_i \approx 0$ and for $z_i \approx 8.8\,\mu m$ $\alpha_i^{max} \approx \Omega_i$. This means that for thin crystals or molecules close to the surface the birefringence of p-terphenyl can be neglected. The black trace shows all possible α_i^{max} values for molecules with $\Omega_i = +30°$ located at different depths z_i. Within 4.4 μm, α_i^{max} takes all values between $+|\Omega_i|$ and $-|\Omega_i|$ (for $|\Omega_i| <$ 45°). Figure adapted from Ref. 2.

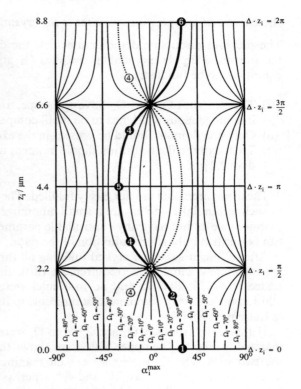

imum fluorescence is obtained for a polarization parallel to the b'-axis. Moving deeper again into the crystal α_i^{max} changes the sign and at position ⑤ becomes equal to $-|\Omega_i|$. Now α_i^{max} increases back toward the value of Ω_i, which is reached at position ⑥ and the next period restarts. For any trace, α_i^{max} takes, within 4.4 μm, all values between $+|\Omega_i|$ and $-|\Omega_i|$ (for $|\Omega_i| < 45°$). If $z_i \cdot \Delta > \pi/2$, it is no longer possible to determine the sign of Ω_i from (eq. 2). The condition to uniquely determine Ω_i is clarified in Fig. 4: molecules at four positions (④, ④, ④, ④) feature the same angle α_i^{max} and lead to the same value of Ω_i, instead of $(-, +, +, -)\,|\Omega_i|$. To measure the true value of Ω_i, the p-terphenyl flake has to be thinner than $z^{max} = 2.2\,\mu m$, otherwise only its absolute value can be obtained.

Observation of domains

Let us consider molecule X in Fig. 1, located in domain α, with a transition dipole moment of $-27°$ relative to the b'-axis. Molecule Y located in domain β has an orientation of $+27°$. Both molecules have the same environment and therefore belong to the same optical site. However their transition dipole moment direction is different. If both domains are present in the probing volume a histogram of the distribution of the Ω_i values has two peaks symmetrically arranged relative to 0° (see [1]). If only one domain is present, just one peak is observed.

Assignment of the spectroscopic sites to the crystallographic sites

The previous discussion demonstrates that the determination of the orientation of the transition dipole moment in the sites O_1 and O_2 requires the following three conditions to be fulfilled:

(i) The two sites O_1 and O_2 have to be spectrally separated clearly, that is, their inhomogeneous width has to be small compared to their spectral separation.
(ii) Only one domain should be present in the excitation volume.
(iii) To determine the transition dipole moment uniquely the crystal must be thinner than 2.2 μm.

These requirements are not easily fulfilled. The handling and stress-free mounting of such a thin flake – to ensure a small inhomogeneous width – is very critical. Furthermore we do not have any controllable parameter to fulfill requirement (ii), and it can be verified only after evaluation of the data.

After a longer search, a crystal fulfilling all three requirements was found. It was approximately 2 μm thick (requirement (iii)), the inhomogeneous width was estimated to be 15–20 GHz and no molecules were found between O_1 and O_2. This allowed the unique assignment of a molecule to its site, based on its spectral position (requirement (i)) [3].

31 molecules from site O_1 and 29 from O_2 were investigated. The results are summarized in Fig. 5. The upper histogram shows the measured distribution of α_i^{max} in the two sites. The O_1-molecules exhibit maximal fluorescence for incident polarization angles α_i^{max} between 21° and 49°, whereas the value of α_i^{max} for O_2-molecules is between −23° and −4°. The investigated volume has only one domain, since each site exhibits a unique peak. The lower histogram shows the transition dipole moment orientations Ω_i after the birefringence correction as calculated from (eq. 2). The distributions are narrower and the peaks are shifted to larger values of $|\Omega_i|$. The absolute values $|\Omega_i|$ of the O_1 molecules are larger than those in O_2. Their average values are:

$$\text{for molecules in } O_1: \quad \bar{\Omega}_{O_1} = +39° \quad \text{and}$$

$$\text{for molecules in } O_2: \quad \bar{\Omega}_{O_2} = -28°.$$

The average angle difference between the orientations of the dipole moments in O_1 compared to O_2 is approximately 67°.

The high intersystem crossing rate in the sites O_3 and O_4 prevents single molecule experiments at present. Still, a qualitative understanding of the orientation of the transition dipole moments can be gained using a highly doped crystal. The 10^{-6} mol/mol sample was mounted on a pinhole of 1 mm diameter. The experimental procedure was the same as for single molecules. The wavelength was tuned to the center of the different origins and the fluorescence was measured as a function of the exciting polarization. Fig. 6 summarizes the results of these experiments for all four sites. The birefringence of the very thin crystal (~ 0.7 μm) can be neglected. The maximum fluorescence intensity of site O_1 and O_2 respectively O_3 and O_4 differs by 28°. The

Figure 5. The upper histograms show the distribution of the angles of the incident polarization for maximal fluorescence intensity α_i^{max} of molecules in the O_1 and O_2 site. The lower histograms show the distribution of the orientations of the transition dipole moments Ω_i, as calculated from Eq. (2). The average orientation of the transition dipole moment for O_1 molecules is $\bar{\Omega}_{O_1} = +39°$ whereas for O_2-molecules the value is $\bar{\Omega}_{O_2} = -28°$. Figure adapted from Ref. 3.

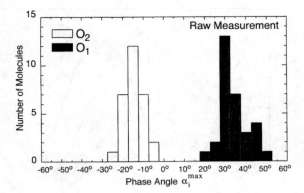

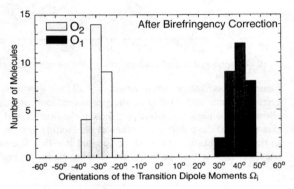

value observed in the single molecule case was 67°. This rather large discrepancy can be caused by an uneven volume ratio of 3 : 1 of the two domains present in the excited volume. However, we can conclude that the angles for maximal fluorescence of O_1 and O_4 are close together and the same holds for O_2 and O_3, where the two pairs have opposite signs. This means that the orientations of the transition dipole moments in the O_1 and O_4 site are tilted away from the b'-axis in the same direction, but opposite to the tilting direction of the sites O_2 and O_3. These data are consistent with bulk data recently obtained from ODMR measurements [9].

There are 24 possibilities to assign the four spectroscopic sites O_1, O_2, O_3, and O_4 to the crystallographic locations M_1, M_2, M_3, and M_4. With the help of bulk measurements the choice can be reduced to 8 possibilities, the ones for which O_1, O_4 (and O_2, O_3) have similar orientations. From these only two are compatible with the single-molecule experiments: comparing the average angles $\bar{\Omega}_{O_1}$ and $\bar{\Omega}_{O_2}$ with Table 1 we see that O_1 molecules are in position M_2 or M_3, depending which domain was investigated in the experiments. The positions M_1 or M_4 are occupied by O_2 molecules (M_1 if O_1 occupies M_3, else M_4). O_4 can now occupy M_1 or M_4 (M_4 if O_1 occupies M_3) and the O_3 molecules will sit in the remaining position. Table 2 summarizes the two most probable assignments among the eight mentioned above. Since the investigated domain remains unknown (α or β), a complete assignment cannot be given.

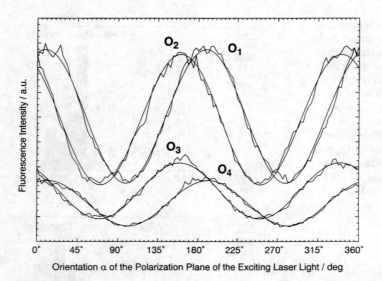

Figure 6. Polarization measurements for all four sites in a thin ($\approx 0.7\,\mu$m), highly doped crystal. The excited volume is $5.5\,10^5\,\mu$m^3, the pentacene concentration is 10^{-6} mol/mol. The small thickness allows one to neglect the birefringence of the host crystal. The laser was tuned in the center of the given site and the polarization plane of the exciting light was scanned, as for the single molecule case. In this representation the b'-axis is located at $-10°$. The orientation of transition dipole moment of site O_1 is close to the one of O_4, the same is true for O_2 and O_3. Figure adapted from Ref. 3.

Table 2. Assignment of the spectroscopic sites to the crystallographic positions. Eight possibilities are selected from the bulk measurements. Only two (* marked) are compatible with the single-molecule results.

M_1	M_2	M_3	M_4	assignment bulk/SMS
O_1	O_4	O_2	O_3	+/−
O_1	O_4	O_3	O_2	+/−
O_2	O_3	O_1	O_4	+/+ domain (β)*
O_2	O_3	O_4	O_1	+/−
O_3	O_2	O_1	O_4	+/−
O_3	O_2	O_4	O_1	+/−
O_4	O_1	O_2	O_3	+/−
O_4	O_1	O_3	O_2	+/+ domain (α)*
$-27°$	$-37°$	$37°$	$27°$	pentacene transition dipole moment

Determination of the depth of a single molecule relative to the surface

The birefringence changes the polarization of the exciting laser light as a function of the penetration depth into the crystal. The modulation depth m_i of the fluorescence signal is strongly dependent on the position z_i. Knowing the orientation Ω_i, the modulation m_i and the difference between the effective refraction indexes for the exciting light in the crystal ($n_{b'} - n_{a'} = 0.067 \pm 0.008$, $\lambda_{\text{vac}} \approx 592$ nm, 300 K [7]), it is

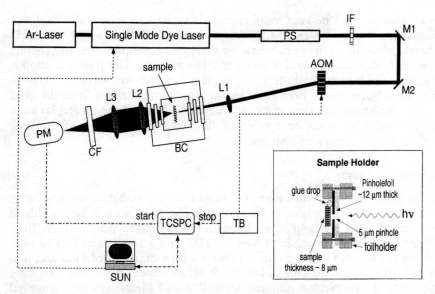

Figure 7. Experimental setup for the measurement of the fluorescence lifetime of single pentacene molecules in a *p*-terphenyl matrix according to Ref. 10. The radiation of a single-mode dye laser with a bandwidth of about 2 MHz passed through a power stabilizer (PS), an interference filter (IF) and an acousto-optical modulator (AOM). The AOM driven by a programmable time base (TB) generated pulses of 9 ns FWHM at a repetition rate of 1 MHz. The sample was glued to a pinhole and mounted in a bath cryostat. A set of lenses (L2, L3) imaged the emerging fluorescence onto a photomultiplier (PM). The cutoff filter CF suppressed the exciting laser radiation. The photomultiplier signal was fed into a time-correlated single-photon counting system (TCSPC). To measure the prompt response of the apparatus, the sample was removed and the cutoff filter CF was replaced by a neutral density filter. The sample was located in the bath cryostat (BC).

possible to calculate the location z_i of the emitting chromophore uniquely using the following relation [2]:

$$z_i = \frac{\lambda_{\mathrm{vac}}}{2\pi \cdot (n_{b'} - n_{a'})} \arcsin \sqrt{\frac{1 - m_i^2}{\sin^2(2\Omega_i)}} \qquad 0 < |\Omega_i| < \frac{\pi}{2} \qquad (3)$$

However, the data has to be interpreted carefully. In Fig. 4 we can see that a small change in the orientation of $\alpha_i^{\max}$ and Ω_i corresponds to a large change in z_i (e.g. from position ① to ②). Hence the error of Ω_i is amplified in the calculation of the depth z_i.

1.3.3 Fluorescence lifetime

The lifetime of the first excited state of single pentacene molecules in *p*-terphenyl was measured by time correlated single photon counting (TCSPC) [10]. The expected lifetime of the S_1-state of pentacene is about 20 ns, and it was necessary to optimize the excitation laser pulse duration carefully. While short pulses are advantageous, the spectral linewidth must be kept small enough in order to resolve individual molecular resonances, which can exhibit a lifetime limited FWHM as low as 8 MHz [11]. The best achievable compromise between these two conditions was to generate 9 ns wide pulses with a transform limited width. The exciting laser pulse duration is then of the same order of magnitude as the expected molecular lifetime. The measured exponential fluorescence decay will incorporate contributions from the laser pulse duration, the apparatus time response and the lifetime of the molecular state itself. Deconvolution [12] must be applied to eliminate the influences of the laser pulse duration and the apparatus response.

1.3.3.1 Experimental setup

Transform limited pulses of a duration of 9 ns with a very stable center frequency were generated as follows: A tunable single mode dye laser with a bandwidth of 2 MHz was pulsed with the help of an acousto-optical modulator. For the fluorescence lifetime experiments pulses of 9 ns FWHM and a separation of 1 μs were used. The laser pulses were focused onto a pinhole with a diameter of 5 μm and illuminated about 100 μm^3 of the sample crystal. A set of lenses imaged the emerging fluorescence onto a fast photomultiplier. Optical cutoff filters removed scattered laser light from fluorescence. Recording times up to 1200 s were necessary to obtain reliable statistics for the fluorescence decay curves. The laser had to be stabilized onto a fixed frequency during the full measurement time.

1.3.3.2 Data analysis

Experimental decay curves measured after a pulsed excitation are affected by the limited frequency response of the detection system and the width of the exciting light pulse. Assuming a linear response for all contributions, the measured decay curve $f(t)$ is given by the convolution integral

$$f(t) = e(t) \otimes h(t). \tag{4}$$

where $e(t) = l(t) \otimes a(t)$ represents the effective excitation function formed by the pulse shape $l(t)$ and the apparatus function $a(t)$, and $h(t)$ is the desired response function of the fluorescence system. If the width of the effective pulse shape is very small compared to the time constant of the fluorescence decay, the excitation pulse does not distort the decay curve very much, so that $e(t)$ can be approximated by a δ-function and $f(t) \approx h(t)$. In our case, the fluorescence lifetime and the exciting laser pulse width become comparable. The effective pulse shape $e(t)$ can be measured directly if the fluorescent sample is replaced by a photodetector which measures the time-dependent intensity of the exciting pulse. Since the detector is placed at the sample position, the light pulse passes through the same setup as during a real measurement. Hence the effective excitation pulse, that is the apparatus function convoluted with the exciting pulse width, is measured. In Fourier space, a convolution transforms into a simple multiplication and deconvolution can in principle easily be performed by dividing the Fourier transformed functions $F(\omega)$ and $E(\omega)$ (of $f(t)$

and $e(t)$). In practice, however, the error properties in Fourier space need to be considered carefully [12].

For lifetime measurements, the typical shape of the true fluorescence decay can be approximated by one or more exponential decay curves.

$$h(t) = \sum_{s=0}^{M} a_s \cdot \exp\left[-\frac{t+b}{\tau_s}\right] \tag{5}$$

a_s denotes the relative contribution of the individual terms and τ_s their time constants. The additional parameter b accounts for experimental time delays between the excitation pulse and the beginning of the exponential part of the decay curve. The goal of the analysis is the determination of the free parameters a_s, τ_s and b in Eq. (5). This problem was solved [12] by transforming the expansion (5) into Fourier space. The free parameters are then obtained by a nonlinear least-squares fit to $H(\omega)$ in Fourier space.

1.3.3.3 Measurements and results

To set up the lifetime experiment, fluorescence excitation spectra were recorded using pulses of 90 ns duration corresponding to a spectral bandwidth of 15 MHz at a repetition rate of 1 MHz. The bandwidth was limited by the pulses rise and fall times, the pulse shapes and the frequency jitter of the laser. Fig. 8(a) shows a typical spectrum with four individual molecular resonances A, B, C and D. On average, the lines are about 25 MHz wide as compared to a homogeneous linewidth of about 8 MHz measured by earlier experiments using cw radiation and lower excitation energies. The best fit of the absorption profile of molecule C was obtained using a Lorentzian profile with a FWHM of 27 MHz.

The spectrum in Fig. 8(b), shows the effect of a pulse width reduction to 9 ns. This spectrum represents the conditions under which the actual fluorescence lifetime measurements were performed. Due to the shorter pulse duration, the spectral width of the excitation light severely broadens the molecular resonances. In this case typical linewidths of 75 MHz were observed. Fitting again the profile of molecule C, the best results were obtained using a Gaussian lineshape with a FWHM of 73 MHz. The resonances of molecules A and B are separated by 60 MHz. In spectrum (a) both resonances are clearly resolved. At a pulse duration of 9 ns both peaks cannot be distinguished any more so that both molecules are not suitable for the individual fluorescence lifetime measurements (spectrum 8b). Molecules C and D are ideal candidates, since they are clearly resolved and do not overlap even at 9 ns pulse duration.

Fig. 9 shows the exponential decay of the fluorescence of molecule C excited by a single 9 ns laser pulse at its resonance frequency. The number of photons counted was averaged over the repetition interval of the laser and normalized to unity at its maximum. The solid curve represents the result of the deconvolution procedure described earlier using a single exponential fit function and a constant background. The lifetime obtained from this fit was 23.9 ± 1 ns for molecule C and 24.5 ± 1 ns for molecule D. The introduction of an additional exponential function resulted in a

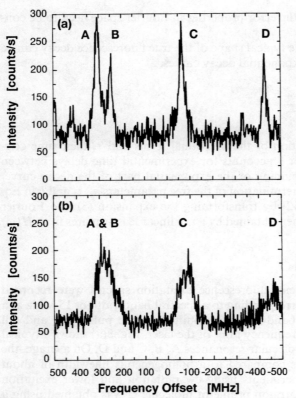

Figure 8. Spectra of single pentacene molecules in a *p*-terphenyl matrix using (a) 90 ns and (b) 9 ns pulses. In spectrum (a) the individual resonances feature an average width of 23 MHz and are clearly resolved. In spectrum (b) the wider excitation bandwidth broadens the absorption profiles, resulting in more than 70 MHz wide resonances. Here molecules A and B cannot be distinguished anymore. Adapted from Ref. 10.

somewhat better fit and in a slight reduction of the fast fluorescence lifetime by 1–2 ns. The time constant of the second exponential evaluated to 2 µs. Table 3 summarizes the results obtained for four investigated molecules. The results of both fitting methods, single and double exponential decays, are given for comparison. Currently no interpretation of the observed long-living decay can be given. However, emission from the triplet state with a lifetime of about 20 µs may be excluded, since the difference in the two time constants seems to be too large.

Previous lifetime measurements on higher concentrated samples [6, 13, 14] resulted in time constants of 21.7–24.5 ns and agree very well with the presented results on individual molecules. Although only four different molecules were investigated, all time constants fall within the experimental accuracy interval and the statistical variation is very low. The average of the four time constants is 24 ns (single exponential fit) with a standard variation of 0.5 ns or 2%. As expected, the lifetime of the excited molecule is not very sensitive to its local nano-environment.

1.3.4 External electric fields and Stark effect

The energy levels of the molecular states are shifted in an electric field, resulting in a spectral shift of the molecular resonances. Single molecule spectroscopy allows one

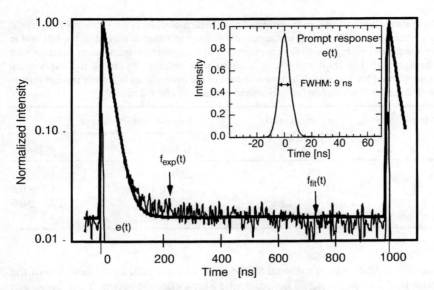

Figure 9. Temporal decay of the fluorescence of molecule C excited by a 9 ns laser pulse at its resonance frequency [10]. The solid curve represents the deconvolution in the Fourier space using a single exponential function as described in the text. In this fit, the background was assumed to be constant. The small inset illustrates the time profile of the exciting laser pulse.

to detect changes in the resonance frequencies in the order of 1 MHz and offers a very sensitive method to measure minute spectral shifts even under weak electric fields. In this section we will concentrate on static electric fields. The so called AC Stark effect, induced by a strong pump laser field, is discussed in chapter 1.2.4.4.

1.3.4.1 Theoretical overview

The energy $W(\vec{E})$ of a level in an electric field $\vec{E}$ can be approximated by a Taylor series

$$W(\vec{E}) = W(0) - \vec{\mu} \cdot \vec{E} - \tfrac{1}{2}((\vec{E} \cdot \hat{\alpha}) \cdot \vec{E}) - \tfrac{1}{6}((\vec{E} \cdot \hat{\beta}) \cdot \vec{E}) \cdot \vec{E} - \cdots \qquad (6)$$

where $\vec{\mu}$ denotes the permanent dipole moment, $\hat{\alpha}$ the polarizability, and $\hat{\beta}$ the hyperpolarizability of the specific molecular state. Isolated, centrosymmetric molecules, like pentacene, have no permanent dipole moment. Therefore, the odd terms of the Taylor series like $\vec{\mu}$ and $\hat{\beta}$ vanish. In these cases the lowest order of the Stark shift is quadratic in $\vec{E}$. Molecular resonances always depend on two energy levels, in our case the ground state S_0 and the exited state S_1. Both states are Stark-shifted in the presence of an electric field, but their Taylor coefficients $\vec{\mu}$, $\hat{\alpha}$, and $\hat{\beta}$ will be different. The spectral shift of a molecular transition will then depend on the Stark shifts of both levels

$$h \cdot \Delta v = [W_{S_1}(\vec{E}) - W_{S_1}(0)] - [W_{S_0}(\vec{E}) - W_{S_0}(0)] = \Delta W_{S_1}(\vec{E}) - \Delta W_{S_0}(\vec{E}).$$
$$(7)$$

Table 3. Fluorescence lifetimes of four investigated pentacene molecules in a p-terphenyl host crystal [10]. The data were analyzed according to the deconvolution algorithm described in the text and in Ref. 12. The second column represents the time constant of a single exponential decay function with a constant background. The third and fourth columns were obtained by fitting two exponential decay functions without background. The time constant of the faster decay is slightly smaller than in the first case. The longer contribution features a time constant of about 2 µs.

Wavelength [nm]	one exponential decay τ [ns]	two exponential decays	
		τ_1 [ns]	τ_2 [µs]
592.323 (D)	24.5 ± 1	22.9 ± 1	1.7 ± 0.3
592.323 (C)	23.9 ± 1	22.5 ± 1	1.7 ± 0.3
592.374	23.4 ± 1	22.6 ± 1	2.1 ± 0.3
592.375	24.2 ± 1	23.5 ± 1	1.7 ± 0.3

The linear Stark shift is proportional to the local electric field at the position of the molecule and to $\Delta\vec{\mu} = \vec{\mu}_{S_1} - \vec{\mu}_{S_0}$. The quadratic Stark effect involves $\Delta\hat{\alpha} = \hat{\alpha}_{S_1} - \hat{\alpha}_{S_0}$. A solid may feature local electric fields, even if there is no external field $\vec{E}_{ext}$ applied. This internal electric field $\vec{E}_{int}$ is dominated by the local charge distribution of the nearest environment. Internal and external electric field together form the local electric field at the position of the molecule.

1.3.4.2 Linear Stark effect of single terrylene molecules in polyethylene

Terrylene, a higher homolog of perylene, was embedded in polyethylene [15]. It features a conveniently located strong absorption band at 569 nm approximately 10 nm in width. Hole-burning experiments showed a very low burning efficiency. Long exposure times and high burning intensities resulted in very shallow holes only. These characteristics, high absorption, high fluorescence yield, and a low burning efficiency, make terrylene in polyethylene an excellent candidate for single-molecule spectroscopy.

Experimental setup

A small piece of terrylene doped polyethylene sheet (high density, crystallinity between 60% and 80%) was optically contacted to the end of a single-mode optical fibre using an index-adapting oil. No adjustments were necessary, since the illuminated area was determined by the diameter of the fibre core (about 4 µm). The end of the fibre (together with the sample) was mounted in a Stark cell. The electrodes consisted of two small pieces of plane parallel glass with a silver coating on their back faces. Special spacers fixed the distance between the two glass plates to 0.24 mm. Each plate was 0.13 mm thick, resulting in a total distance of 0.6 mm between the electrodes. The homogeneous electric field was applied in the plane of the sample and perpendicular to the direction of the exciting laser beam. With voltages up to 100 V, only rather low electric field strengths could be achieved. The rather large dimensions of the Stark cell were especially advantageous with respect to a high uniformity of the electric field throughout the illuminated sample.

The optical setup of the Stark experiment is comparable to a hole-burning setup described in [16]. The fiber setup for laser excitation and fluorescence collection has been described in previous chapters. Since the fiber did not preserve the polarization, the orientation of the exciting laser light with respect to the electric field was not known.

Measurements and results

When scanning 30 GHz frequency intervals, single molecule signals were found as very narrow lines at the red side of the maximum of the excitation spectrum. The investigated region covered a wavelength interval from 571 to 575 nm. Even though the recording power was only about $0.5 \, mW/cm^2$, light induced frequency shifts of the molecular absorption line were observed. The line widths of the investigated molecules varied from 60 to 150 MHz. Experiments with further attenuated laser power showed no decrease of the line widths. Power broadening of the single molecule lines was negligible. The observed line widths result mainly from optical dephasing or spectral diffusion during the measurement period (a few of seconds). The rather large variation of the line width for different molecules may be attributed to the inhomogeneity of the polyethylene matrix: polyethylene consists of crystalline as well as disordered regions.

Under weak electric fields (voltage lower than 100 V between the electrodes 0.6 mm apart) single-molecule lines shifted linearly by different amounts and signs. No broadening of the line width was observed. In fact, only the center frequency of the line shifted proportionally to the applied electric field. The line shape, its height and width, remained unchanged. Even though terrylene is centrosymmetric, all investigated molecules showed a nearly perfect linear dependence of the spectral shift versus the applied electric field (see Fig. 10) [15].

The molecular symmetry is broken by the interaction with the matrix. The slope of the spectral shift versus the electric field mainly depends on $\Delta\vec{\mu}$, the change of the local dipole moment, which itself depends on the respective orientations of the applied field, the orientation of the molecule, and its local nano-environment. More precisely, the component $\Delta\mu_E$ of $\Delta\vec{\mu}$ along the local electric field relates to the slope $d\nu/dE$ by the equation:

$$h \cdot \Delta\nu = -\Delta\mu_E \cdot f_e \cdot E. \tag{8}$$

For a qualitative evaluation of $\Delta\mu_E$ a uniform electric field with a local field factor of $f_e = 1.5$ was assumed. This simplification neglects the influences of the glass electrodes and the optical fibre. Using this model, approximate values of $\Delta\mu_E$ can be calculated. The histogram of Fig. 11 summarizes the results for 60 molecules.

The average value is about 0.44 D. The histogram also shows four lines with a surprisingly large $\Delta\mu_E$ (1.8 D), corresponding probably to strongly distorted terrylene molecules.

This kind of line shape preserving shift of the resonance frequency as a function of the applied electric field was expected. In the view of the presented Stark effect of

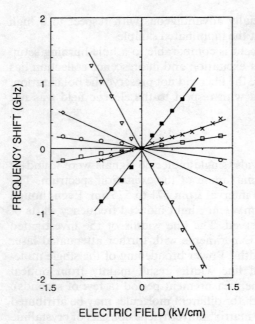

Figure 10. Dependence of the spectral shift of the center frequency of individual molecule lines on the applied external electrical field according to Ref. 15. The solid lines represent linear fits to the observed shifts. The magnitude and sign of the shifts vary from molecule to molecule depending on the orientation of the field with respect to the molecular symmetry axis.

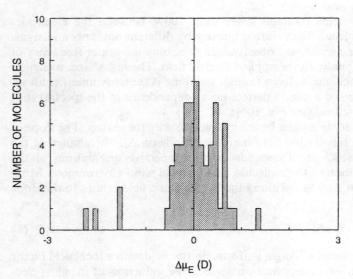

Figure 11. Statistical distribution of the values of $\Delta\mu_E$ for 60 terrylene molecules in a polyethylene matrix [15]. The distribution is centered around 0, but not symmetric. Thus, a small average value $\Delta\mu_E = 0.44$ D remains. Some of the terrylene molecules exhibit a very strong change of the dipole moment $\Delta\mu_E \approx$ 1.7 D, which might result from strong distortions.

single molecules, the broadening of spectral holes under electric fields [17] is easily understood: since many individual molecules contribute to hole-burning, the behavior of a spectral hole is only a statistical effect. Each individual contributing molecule will shift according to the presented measurements, without larger variation of its line shape. But each of these molecules has a different orientation and will

Figure 12. Stark effect of perylene in *n*-nonane. See text for details. Adapted from Ref. 19.

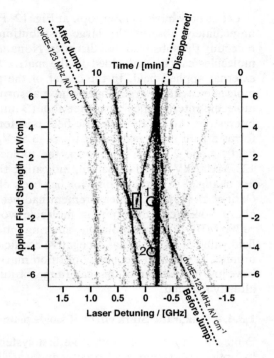

therefore have a unique slope of its spectral shift versus the electric field. On average, the individual shifts of all contributing molecules result in the observed broadening of the spectral hole.

1.3.4.3 Perylene in nonane

In the system perylene in *n*-nonane, the main 0–0 absorption band at liquid helium temperatures is located around 443.8 nm with a few weaker sites. To work in this spectral region a cw Ti : Sapphire autoscan laser is frequency-doubled by means of a LiIO$_3$-crystal [18]. For the following experiments a tiny drop of a 10^{-7} to 10^{-8} mol l^{-1} solution of perylene in *n*-nonane is squeezed between two thin glass plates, covered with a conducting sheet, and placed after a 5 μm pinhole. The spacing between the electrodes is 300 μm. A microscope objective immersed in the liquid helium is used to collect the fluorescence [19]. An overview of the results obtained for this system is given in Fig. 12.

In a typical experiment, the frequency of the laser is slowly increased in steps while the electric field is scanned at a fast rate ($\sim$45 scans/s). Fig. 12 reflects the way the data were recorded: the *x*-axis represents the laser frequency whereas the *y*-axis represents the applied electric field strength. The *x*-axis can also be interpreted as a time axis of the experiment running from right to left. It took 2.1 s to complete the 100 electric field scans for one frequency position. One can see several single-molecule traces shifting linearly as a function of the applied electric field.

Let us now have a closer look at Fig. 12. From the single molecule traces, one can immediately observe the large distribution of Stark coefficients in this system, reflecting the substitutional disorder. Nonane is a Shpol'skii system and the perylene molecules can be embedded in the matrix in different orientations relative to the external applied field. In the center of the plot around -0.2 GHz there are some molecules, with a large slope, whose absorption lines overlap. The behavior of the molecule entering the figure after about 3 minutes at a field of -6 kV/cm is especially interesting. At the time-electric field position labeled with circle 1, it suddenly performs a field jump equivalent to ≈ -3.5 kV/cm (along the direction of the applied field), circle 2. After the jump, the molecule has the same Stark coefficient of 123 MHz/(kV cm^{-1}) as before, indicating that the orientation of the molecule has not changed. A similar behavior has been observed on terrylene in polyethylene by Orrit et al. [15]. If the measurement had been performed with a fixed Stark field, at about -3 kV/cm, the jumping molecule would have appeared twice, separated by ≈ 300 MHz. Another molecule, entering into the figure after 5 minutes at an electric field value of 6 kV/cm has a negative coefficient and it disappears completely after 7 minutes at the field-frequency position marked by the square. This molecule might have just jumped out of the scan range, or jumped to the red, in a position which was already scanned.

1.3.4.4 Quadratic Stark effect of single pentacene molecules in *p*-terphenyl

Pentacene in *p*-terphenyl was the first system where single molecules were detected by optical excitation spectroscopy and was also the first system in which the effect of external electric fields on single molecules [20] was studied.

Experimental setup

The pinhole setup previously described was modified for Stark shift experiments. Fig. 13 shows a close-up and indicates the proper dimensions of the components. A

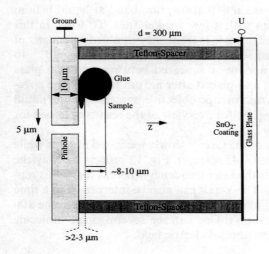

Figure 13. Close-up view of the Stark cell used in Ref. 10. A 5 μm hole in a 10 μm thick stainless steel disk served as aperture. At the same time the steel disk was used as ground electrode for the electrical field. The anode separated by spacers of 300 μm was made out of a glass plate coated with SnO_2. A DC-Voltage of 0–1 kV resulted in an electrical field of 0 to 3.3 MV/m in the e_z-direction.

Figure 14. Timing of the Stark effect measurements. To overcome the long-term laser drift, two measurements at $+E$ and $-E$ were made in combination with two calibration measurements without electrical field at the same laser frequency. The calibration spectra allowed one to quantify potential laser drifts and to correct the Stark spectra accordingly.

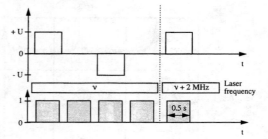

5 μm hole in a 10 μm thick stainless steel disk served as aperture. At the same time, the disk piece was used as ground electrode of the electric field. A SnO_2-coated glass plate formed the anode, which was separated from the cathode by teflon-spacers of 300 μm thickness. A DC voltage between 0 and 1 kV could be applied to the anode and resulted in an electric field of 0 to 3.3 MV/m in the $-z$-direction. A drop of glue inhibited a direct contact between the sample and the ground electrode and guaranteed a good electrical insulation. The distance between the pinhole plate and the sample varied between 2–3 μm for different setups. The electric field distribution between the two electrodes can be approximated by the field of a parallel capacitor, since radial components can be neglected at distances greater than 2.5 μm from the pinhole [2].

Measurement procedure

The fluorescence spectra of single pentacene molecules were obtained by frequency scans of 200 MHz divided into 2 MHz steps. The long-term laser drift during the experiment significantly affected the absolute wavelength calibration of the measurements. To overcome this problem, the following data recording scheme was used (Fig. 14): At each frequency position the electric field was switched between four different values $+E$, 0, $-E$ and 0. For each electric field the number of arriving photons was counted over a period of about 0.5 s. The simultaneously recorded four spectra represent the Stark shifts for positive and negative electric field strengths, as well as two calibration spectra without field. The calibration spectra made it possible to quantify the potential laser drift and to correct the Stark spectra, accordingly. Following this procedure, measurements at different electric field strengths (0–3.3 MV/m) were performed for each individual pentacene molecule.

Results

22 pentacene molecules of the O_1-site of the $S_1 \leftarrow S_0$ transition at 592.32 nm were investigated [2]. As an example, Fig. 15 shows the Stark shift of four individual pentacene molecules at a voltage of +29 kV/cm and −29 kV/cm. Two spectra at 0 V are included as reference. Molecule A clearly shows a shift Δv towards smaller frequencies. The direction and magnitude of this shift are independent of the polarity of the applied voltage. This is a first indication of a quadratic Stark effect, which has already been found for the full inhomogeneous bands of pentacene and tetracene in

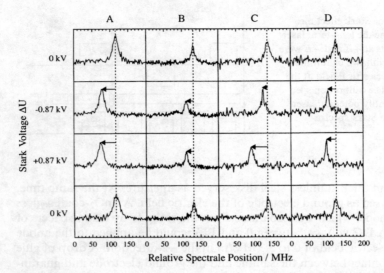

Figure 15. Stark shift of individual pentacene molecules [20]. The molecular resonances are shifted towards smaller frequencies, independent from the sign of the electrical field ($U = 0.87$ kV). The magnitude of the frequency shift varies from molecule to molecule and depends for molecule C additionally on the sign of the electrical field.

p-terphenyl by Meyling et al. [21]. Molecules B and D show approximately the same behavior, where the magnitude of the shift is somewhat smaller. In contrast, Molecule C features a strongly asymmetric shift depending on the sign of the applied voltage.

Fig. 16 plots the experimental results (circles) and best fits (solid curve) of the frequency shifts versus the magnitude of the electric field for the same molecules of Fig. 15. Clearly, the previous assumption is verified: all molecules show a quadratic dependence of the frequency shift on the external electric field. Molecules A, B, and D shift symmetrically with respect to the zero value of the applied voltage, whereas molecule C shows a significant asymmetry, which can be modeled only by an additional linear contribution.

The solid curves of Fig. 16 were obtained by fitting a second-order polynomial to the experimental results giving $\Delta\alpha$ and $\Delta\mu$. $\Delta\alpha$ is positive for all molecules, indicating that the polarizability of the excited state is always larger than in the ground state. The precise value varies considerably from molecule to molecule. The mean value of $\Delta\alpha$ for all observed molecules is

$$\overline{\Delta\alpha} = (47 \pm 7) \times 10^{-40}\,\mathrm{Fm}^2. \tag{9}$$

Earlier measurements on a bulk sample reported a shift of the complete inhomogeneous band of $\overline{\Delta\alpha_{\mathrm{bulk}}} = (86 \pm 4) \times 10^{-40}\,\mathrm{Fm}^2$ (for $f = 1$) [21], which was later corrected to $\overline{\Delta\alpha_{\mathrm{bulk}}} = (23 \pm 2) \times 10^{-40}\,\mathrm{Fm}^2$ (for $f = 1$) [22]. Even though pentacene

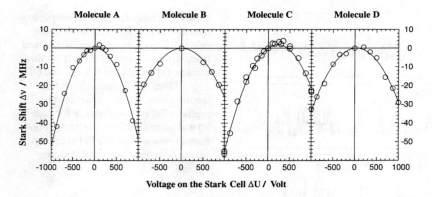

Figure 16. Experimental (circles) and fitted (lines) Stark shifts of four individual pentacene molecules as a function of the applied electrical field [20]. Clearly a quadratic dependence of varying magnitude dominates. In case of molecule C an additional linear term results in an asymmetric behavior and a shifted Stark parabola.

features only a very small burning efficiency, Moerner et al. were able to measure its Stark shift using spectral hole-burning. In [23] they report a value of $\overline{\Delta\alpha_{\text{bulk}}} = 75 \times 10^{-40}\,\text{Fm}^2$ (for $f = 1$). The precise value of $\Delta\alpha$ is difficult to assign, since it strongly depends on the correction model for the internal electric field $E_{\text{int}} = f \cdot E_{\text{ext}}$ used. The anisotropic Lorentzian correction gives a factor $f = 1.92$ in the direction of the pentacene long axis [22].

Many of the investigated molecules exhibit a small linear contribution to the quadratic Stark shift, which indicates deviation from the centrosymmetric geometry of the molecule. In these cases, an internal electric field $\vec{E}_{\text{int}}$ may induce an additional dipole moment $\vec{\mu}_{\text{ind}}$. Compared to an external Stark field of 3.3 MV/m, the internal electric fields are significantly smaller.

Meyerling et al. [21, 22] were not able to measure linear contributions to the quadratic Stark shift since their experiment was not sensitive enough. Here, single molecule spectroscopy really reveals its full potential. The individual nature of the measurement technique allows one to detect the internal electric fields in the local environment of each molecule, whereas bulk experiments using samples of higher concentration cannot measure the vanishing averaged field components.

1.3.4.5 Stark effect in the optical near-field

Optical near-field microscopy (SNOM and NSOM [24, 25]) is a novel technique with a spatial resolution power well beyond the diffraction limit. It was applied to image the fluorescence of single molecules at room temperature [26, 27] (see Chapter 2). Typical spectral linewidths reported were around 20 000 GHz [27], due to the room temperature operation. At liquid helium temperatures the linewidths are reduced to about 10–100 MHz, which allows study of the special effects on single molecular resonators. Moerner et al. [28, 29] combined the SNOM-technique with Single Mol-

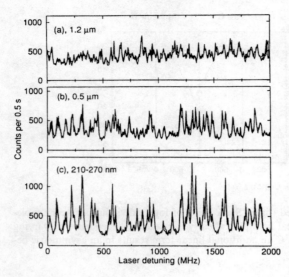

Figure 17. Near-field spectroscopy. Statistical fine structure and single-molecule features upon approach to the near field. Approximate distances from the surface: (a) 1.2 μm, (b) 0.5 μm, and (c) 210–270 nm. Each panel shows two spectra taken ≈ 5 min apart to show reproducibility. The distance for case (c) was estimated from the background fluorescence increase (BFI) approach curve. Zero detuning = 592.066 nm, $V_T = 0$.

ecule Spectroscopy (SMS) at cryogenic temperatures to study individual pentacene molecules at the surface of a p-terphenyl crystal in the optical near-field (Near-Field Single-Molecule Spectroscopy, NF-SMS).

Experimental setup

The optical apertures used in near-field microscopy are usually prepared by pulling a heated optical fibre until it breaks [25]. The sides of the tips are coated with aluminum. The typical diameters of the apertures produced by this technique are 60 ± 10 nm, which is about one tenth of the optical wavelength. The transmission of such a tip ranges from 10^{-5} to 10^{-6}. The near-field tip is mounted on a xyz-piezo-electric (PZT) tube scanner to control the fine approach (z) to the surface and the lateral dithering (x, y) of the tip. The coarse z positioning was achieved by a coupled spring and steel plate comparable to the setup described in [30]. The sample was connected to a small glass hemisphere to minimize losses due to internal reflections and mounted in the focus of a paraboloid mirror with a numerical aperture of NA = 0.98. The whole setup, paraboloid mirror, sample, and PZT tube with the fibre tip, was then mounted inside a cryostat and immersed in superfluid Helium at 1.8 K.

Results

Fig. 17 shows three examples of fluorescence spectra measured at different distances from the surface. At 1.2 μm only statistical fine structure can be observed. In this case the illuminated area is very large and many molecules with the same resonance frequency are excited. Their fluorescence overlaps and it is not possible to distinguish individual molecules. Nevertheless, the fine structure spectrum is reproducible, as demonstrated by the two traces taken 5 min. apart. Closer to the surface the single molecule features become more pronounced. The excited volume is now smaller and

Figure 18. Excitation spectra showing three methods to identify molecules close to the tip at 1.1X (sample 2, $d \approx 250$ nm. Upper panel: saturation method: spectra taken with (a) $P_c = 100\,\mu W$, and (b) 25 μW. Middle panel: static Stark shift method, traces labeled by V_T. Bottom panel: Stark shift with transverse dithering method, traces labeled by V_T. The scale is exact for the lowest trace in each panel while the other traces are shifted vertically upward. (0 detuning = 592.067 nm.)

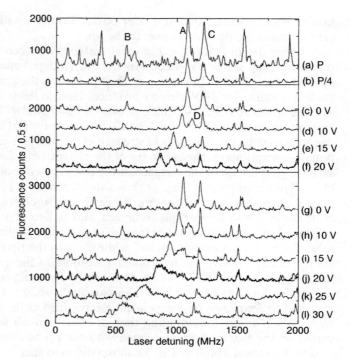

less molecules are excited. Therefore, the probability of spectral overlaps between individual molecules decreases. At 250 nm the fluorescence signal drops between each molecule to a constant background and the intensity of the emitted fluorescence increases. The latter is caused by a more efficient coupling of the exciting light to the molecular resonator. The linewidth of the pentacene molecules is about 10–20 MHz, which agrees very well with other measurements presented in this book.

The approach of the tip to the sample surface was carefully monitored by observing the background fluorescence increase (BFI) [28] and by shear force detection. It was stopped at about 250 nm. The investigated molecules are not necessarily near the surface. Three techniques were used to determine the distance between the tip and the individual molecule: (i) saturation measurements, (ii) static Stark effect, and (iii) static Stark effect with lateral dithering. Fig. 18 illustrates typical measurements using these techniques.

Molecules near the tip are expected to saturate easily. Reducing the excitation power should decrease their signal strength less than in a linear way. The spectrum (a) in Fig. 18, upper graph, was taken at a power $P = 100\,\mu W$ and features strong resonances A, B, and C which seem to be power-breadened. In the lower spectrum (b) (and all other spectra) the laser power was reduced by a factor of 4. The unsaturated features decrease approximately proportional with the laser power, while features A, B, and C decreased only slightly. A systematic analysis resulted in a low power linewidth of 20 ± 2 MHz and 13 ± 2 MHz for molecules A and B. In addition, molecule A saturated at a laser power 5 times smaller than molecule B, indicating that molecule A was closer to the tip.

A better determination of the axial distance between tip and molecule can be obtained using the Stark effect. An electric voltage between 0 and 50 V could be applied to the aluminum coating of the tip. Due to the small distance to the sample surface a voltage of 10 V already produces electric fields up to 10^6 V/cm. Since the average quadratic Stark-coefficient of $-3.3 \, 10^{-8}$ MHz/(V/cm)^2 was known from previous experiments ([20] and 1.3.4.4), the measurement of the spectral shift allows one to determine the electric field strength at the position of the molecule. With additional assumptions on the geometry of the electric field distribution between tip and surface, an estimate of the distance of the investigated molecule from the tip may be given. The central part of Fig. 18 shows the dependence of the molecular resonances on the applied voltage of the tip. While several molecules are not affected at all, molecule A shows a clear quadratic red shift ((c) to (f)). The different behavior is attributed to a different distance from the tip. At a large distance, the variation of the local electric field is not significant, while close to the tip a large electric field affects the molecules strongly. With certain assumptions of the qualitative field distribution [28] a distance of $r = 400 \pm 50$ nm from the tip for molecule A was estimated.

Finally, dithering of the tip with an amplitude of 20 nm and a frequency of 7 kHz was used to generate a time dependent electric field at the locations of the individual molecules. The photon counting time of 0.5 s is much longer than the oscillation period of the electric field associated with the tip position and an averaged signal results. The lower graph of Fig. 18 shows different spectra (g) to (l) taken at increasing electric field strengths. Again, molecule A features a quadratic shift of its central resonance frequency. Using a simple spherical tip and field model, the broadening and Stark shift scale with the square of the distance from the tip and result in an upper bound of 400 nm for molecule A, independent of the actual Stark shift coefficient.

1.3.5 Pressure effect

Single molecules at low temperatures are ideally suited to study the influence of very small external perturbations. Pressure is such an interesting property, since it varies distance and orientation of molecules, which are the parameters determining intermolecular potentials. In fact the primary effect of a small pressure increase is to change in the intermolecular interactions, caused by reducing the average separation of molecules. In the past, properties like location, width, structure, and relative shape of spectroscopic bands, including oscillator strengths and lifetimes, in solids and in solution, have been studied as a function of pressure in the kbar (GPa) region [31]. These very high pressures are needed in conventional spectroscopy in order to observe an effect on the whole inhomogeneously broadened band. They can no longer be regarded as small perturbations since they often lead to irreversible changes in the system. For instance, above 5.5 kbar at 15 K pentacene in *p*-terphenyl features a phase transition from triclinic (four sites) to monoclinic (one site) [32].

The increase of spectral resolution (10^3–10^5) obtained with spectral hole-burning (see e.g. Ref. 33 for a review) allows measurements of pressure effects on homogeneous subsets of molecules in the range of a few hundred hPa to MPa at low temperatures. In such experiments, a spectral hole was burned at a given pressure in the inhomogeneously broadened band of guest molecules in a host material and its spectral position and shape are measured at different pressures. In the case of hydrostatic pressure, the effect was found to be reversible [34], showing a spectral shift and a broadening of the spectral holes. The frequency shift results from a change in the host density, while the hole broadening occurs due to the removal of accidental degeneracy of the transition frequencies of the burned molecules. The broadening therefore provides information on the degree of microscopic disorder in the host material. The greater the pressure change, however, the more difficult becomes the detection of the holes. At a certain value of the pressure change, depending on the sensitivity of the detection method, the hole will become too broad to be detectable. Single molecules, on the other hand, are sensitive to very small pressure changes and do not suffer from line broadening. Furthermore, the distribution of spectral shifts, which causes the broadening in hole burning, can now directly be accessed. The pressure effect on single molecules has been studied for two different kinds of molecules doped in the same host: first it was demonstrated in the classical system pentacene *p*-terphenyl [35] and then a study of terrylene in *p*-terphenyl [36] was performed [37].

1.3.5.1 Experimental

The optical setups for these experiments are very similar to the setups described in the previous chapters. The pentacene study used a pinhole to reduce the excitation volume. The setup consisted of a home-built pressure cell and an external home-built reference cavity, to account for the slow frequency drifts of the commercial dye laser. In the terrylene study a magnet controlled lens, immersed in liquid helium, focused the laser light to a tiny spot and an ellipsoidal mirror collected the fluorescence.

Reference cavity

In these experiments, relative changes in the frequency positions of single molecules are measured over a long period of time. Therefore, the stability of the laser frequency has to be either increased, or monitored and taken into account in the evaluation of the data. The second option was chosen for the pentacene study. A home built, temperature stabilized cavity was actively locked to a commercial, frequency stabilized He–Ne laser (specified frequency drift < 1 MHz/day). During a frequency scan the transmission signal of the dye laser through the cavity is simultaneously recorded, together with the single molecule fluorescence signal from the photomultiplier. The transmission signal of the dye laser consists of equally separated peaks. Any shift in the position of the transmission peaks, for two consecutive scans centered at the same frequency, is due to a dye laser frequency drift and can be corrected [35]. In the terrylene study the pressure shift was measured over a larger range

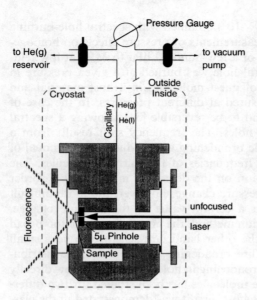

Figure 19. The home-made pressure cell used in Ref. 35 consists of a small volume chamber with two windows. The chamber has a capillary which connects it to a pressure gauge and two valves, one connected to a helium gas reservoir and the other to a vacuum pump. During operation, the cell is immersed in liquid helium. The capillary reaches out of the cryostat, where the gauge and the two valves are located. Liquid helium condenses in the sample chamber, and acts as pressure transmitting medium. The pressure changes in the helium gas phase in the upper part of the capillary, which are controlled by the vacuum pump and the helium reservoir, are transmitted to the sample by the liquid.

as compared to the pentacene case, furthermore, the homogeneous width of terrylene molecules is larger (40 MHz at low power, compared to 8 MHz for pentacene). These two parameters reduce the influences of the laser drift and therefore no special care was taken to monitor it.

The pressure cell

The home-built pressure cell, used for pentacene, consists of a small volume chamber made of copper. The chamber has two windows sealed with indium and a capillary (Fig. 19). The cell fits in the bottom of the cryostat and is connected through the steel capillary to the outside, where a pressure gauge, the helium supply, and the vacuum pump are located. During an experiment the cell is immersed in superfluid liquid helium at a temperature of 1.8 K. Helium condenses in the lower part of the cell and the liquid is used as pressure transmitting medium. Its magnitude is determined by the pressure of the helium gas in the upper part of the capillary, which can be controlled by the gas supply and the vacuum pumps. Hydrostatic pressures from 10 to 10^3 hPa can be applied to the sample.

The pressure cell built by Muller et al. [37] for the terrylene sample is made of stainless steel and is equipped with quartz windows. As mentioned before, it contains both, the focusing and the collecting optics. The cell is connected to a helium gas cylinder and the pressure can be fine-tuned with a mechanical valve. This cell allows the application of pressure changes from 200 hPa to 2500 hPa.

In both setups the pressure of the helium gas is changed at room temperature. It is therefore necessary to wait for a few minutes after each pressure change until the pressure and the temperature of the system have stabilized. It should be pointed out that these studies can only be performed on frequency stable molecules.

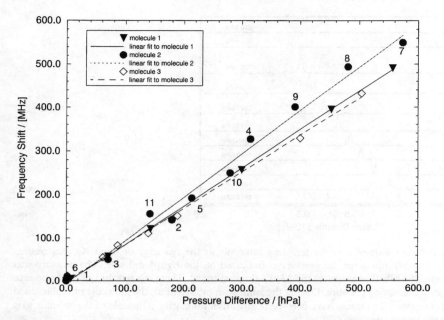

Fiquirs 20. The spectral shifts of the zero-phonon lines are plotted against the changes in the hydrostatic pressure for three different pentacene molecules in the O_1 site of a *p*-terphenyl crystal. Linear fits to the data are also plotted. The numbers indicate the temporal sequence in which the data for molecule M_2 (gray circles) were collected. The process is completely reversible within our experimental accuracy. Adapted from Ref. 35.

1.3.5.2 Results and discussion

Results for pentacene in p-terphenyl

All the investigated pentacene molecules (from sites O_1) featured a linear and reversible red shift without additional broadening upon pressure increase [35]. The measured shifts Δv of the resonance frequency for 3 different pentacene molecules are shown in Fig. 20. Linear fits to the measured data are also plotted. The numbers near the points of molecule M_2 (gray circles) indicate the temporal sequence in which the data were collected. The shifts were completely reversible within the resolution of the experiments. The slope of the linear fits for the five investigated pentacene molecules, spectroscopically located in the red of O_1, varied from -0.74 to -1.0 MHz/hPa with an average of -0.9 ± 0.1 MHz/hPa.

Results for terrylene in p-terphenyl

A similar behavior is observed for terrylene [37]. All 35 investigated terrylene molecules (from the sites X_2, X_3, X_4) featured a linear and reversible red shift. A typical study of a single terrylene molecule is shown in Fig. 21. A molecule spectrally located at 578.82 nm at 1.6 K, was investigated at different pressures from 2000 down to 500 hPa. The baseline of the spectra has been shifted vertically, proportional

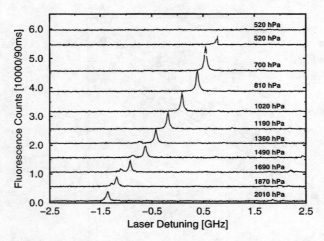

Figure 21. Pressure shift of a single terrylene molecule at the red edge of band X_2 (centered at 578.82 nm). The magnitude of the pressure is indicated on the right-hand side. The pressure was reduced from 2006 hPa (lowest trace) to 523 hPa (highest trace). There is a red shift of the resonance frequency with increasing pressure. During the tenth scan (at 523 hPa), the molecule photobleached and did not reappear. The traces have been vertically offset for clarity. The scale of the ordinate axis is valid for the lowest spectrum [37].

to the pressure change. The molecule resonance shifts to the red with increasing pressure. There is no detectable change of its width with pressure, in contrast to the hole-burning experiments. The observed changes in fluorescence intensity in this experiments are ascribed to mechanical deformation of the optics in the pressure cell. During the tenth scan at 523 hPa the molecule was irreversibly burned away. A sudden drop of the fluorescence signal close to its maximum value is an indication that the process is due to a phototransformation. At the low energy side of the main spectral feature there is a weaker molecule (visible from the second to the fifth scan). It exhibits a smaller pressure shift and disappears after the fifth scan. The magnitude of the pressure shifts for all 35 molecules varied between -0.92 MHz/hPa and -1.48 MHz/hPa with an average of -1.21 MHz/hPa. No correlation between the pressure shift parameter and the spectral position is observed.

Comparison of the two systems

In the framework of solvent shift theories, using a Lennard-Jones type potential for the interaction of a nonpolar dye in a nonpolar matrix, Sesselman et al. [34] developed a simple theory for the interpretation of hole-burning data in the low pressure range (<20 MPa). In this regime, the pressure shift $\Delta\bar{v}$ varies linearly with the solvent shift $\Delta\bar{v}_S$, i.e. the difference between the molecular absorption frequency in the matrix and in vacuum, the local hydrostatic compressibility κ and the pressure change Δp:

$$\Delta\bar{v}(\Delta p) = 2 \cdot \kappa \cdot \Delta\bar{v}_S \cdot \Delta p \qquad (10)$$

Using this model, the calculated compressibility of the host polymers agreed within 10%–20% to the mechanically measured values reported in the literature. We can apply this equation to a single molecule, since it describes only the spectral shift of a homogeneous group of absorbers. The effect of hole-broadening, due to the removal of accidental degeneracy, corresponds, in the single molecule case, to the observation of different pressure shifts for molecules spectrally close together. However, it must be kept in mind that Eq. (10) was derived for amorphous systems, and therefore holds only approximately in this case.

Using Eq. (10), the measured average shift of $-0.9\,\mathrm{MHz/hPa}$ and a solvent shift for a pentacene molecule in the center of O_1 of $-1745\,\mathrm{cm^{-1}}$ (the $S_0(^1A_{1g}) \rightarrow S_1(^1B_{2u})$ absorption of pentacene in a supersonic jet is at $18\,628\,\mathrm{cm^{-1}}$ [38]), a compressibility $\kappa = 0.086 \pm 0.009\,\mathrm{GPa^{-1}}$ is obtained. This value is about half of the compressibility of polymers (like polyethylene, polystyrene) at low temperatures. Sesselman et al. [34] measured a spectral shift for pentacene in polymethyl-methacrylate (PMMA) of $-0.33 \pm 0.02\,\mathrm{cm^{-1}/MPa}$ ($\approx -0.99\,\mathrm{MHz/hPa}$) using hole-burning. The mechanically measured, low temperature compressibility of PMMA is 1.5 times larger than the compressibility we calculated for p-terphenyl. This difference is approximately compensated by the different solvent shifts of pentacene in the two matrices, resulting in a similar pressure shift in Eq. (10).

The absorption spectra of terrylene in the gas phase are not known. Using the compressibility calculated from the pentacene data, the solvent shift of terrylene in p-terphenyl is estimated to be $\Delta \bar{\nu}_S = 2345 \pm 411\,\mathrm{cm^{-1}}$ and therefore its vacuum absorption wavelength should be $509 \pm 11\,\mathrm{nm}$. The larger solvent and pressure shifts found for terrylene as compared to pentacene seem reasonable, since terrylene is bigger and therefore expected to exhibit larger polarizabilities in its ground and excited state.

1.3.6 Fluorescence microscopy

The experimental techniques for single molecule spectroscopy described in the previous chapters differ mainly in the method employed to reduce the excitation volume of the sample (combined with different fluorescence collection methods). This was achieved in four different ways: (i) the laser was focused to a tiny spot on the sample by a lens immersed in liquid helium, (ii) the excitation light was coupled into an optical fiber carrying the sample at its end, (iii) the sample was mounted behind a small aperture (pinhole with typically $5\,\mu m$ diameter). All these methods reduce the excitation area to a few μm^2. The near-field technique (iv) allows investigations beyond the classical diffraction limit: the tapered tip used had a typical diameter in the order of 50–$100\,\mathrm{nm}$.

In the approach described here, microscopy is used to investigate an illuminated area of about $100 \times 100\,\mu m^2$. The small volume to be studied is effectively defined by the optical resolution of the microscope giving as great benefit that many molecules can be investigated simultaneously in parallel. The first experiment performed to spatially localize a single fluorescing molecule in one dimension was realized by translating a focused laser spot, with a width of $\approx 5\,\mu m$, in one direction across the surface of the sample [39, 40]. By taking frequency scans at different positions along a straight line on the sample a three dimensional map of the fluorescence intensity *vs.* *x*-position and frequency could be generated (see Section 1.1.3.2, Fig. 9). Using a truly two-dimensional recording technique, the simultaneous measurement of the lateral (*xy*) position of several individual molecules inside a solid reported [41]. Spatial resolution is one important goal of microscopy, parallelism of data collection is the other. Single molecule microscopy has the capability to investigate many spatially separated molecules in a parallel way, making the experiments more "efficient" and ensuring that all chromophores are investigated under identical macroscopic conditions. This is a highly desirable feature for accessing the distribution of a physical property, e.g. to obtain statistics about the spectral dynamics of a system.

1.3.6.1 The microscope

The standard photomultiplier was replaced by a video camera with additional optical components: A micro-channel plate operating in high gain mode intensified the incoming fluorescence distribution and allowed single-photon detection. A video camera, modified to operate in the non-interlaced mode converted these images into an analog video signal. The camera captured 25 images per second and a fast data processing unit digitized the video signal in real time [41–43]. The different imaging setups developed are described in the following sections.

The first microscope

In the first experiments, performed on the classical system pentacene in *p*-terphenyl, the microscope was still placed at the outside of the cryostat (see Fig. 22, Inset a) [41]. Fig. 23 shows an area of 40 × 40 μm cut out from a few selected images taken at different frequencies. The images are plotted as a two dimensional histogram of the recorded photon intensity. The fluorescence intensity is represented by a grey scale from zero fluorescence (white) to maximum count numbers (black). Molecules are hence dark spots which appear and disappear as the frequency is scanned. Some grey spots which are not correlated with frequency correspond to scattered background light.

   To illustrate the frequency dependence of the signal, the square area in Fig. 23 has been integrated and plotted as a function of frequency in Fig. 24(a). The fitted Lorentzian shape has a width of 14 MHz, consistent with the homogeneous width of a single pentacene molecule. Fig. 24(b) and Fig. 24(c) illustrate the spatial shape of the signal, obtained by integrating over the remaining two dimensions (one spatial dimension and frequency). The gaussian fit allows a localization of the molecule to ±0.5 μm. The FWHM of the intensity distribution was 2.7 μm and 3.8 μm in the *x*- and *y*-direction.

The three-lens microscope

A commercial microscope objective with a numerical aperture of NA = 0.85 and a magnification of 60 was adapted to the limited space inside the bath cryostat. A metal plate with a large pinhole of 50 to 150 μm in diameter held the sample. The

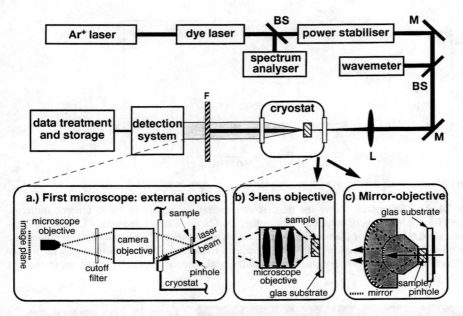

Figure 22. Sketch of the experimental setup of the fluorescence microscope with light source, cryo-stat, and detection system. Three insets show the different optical setups used. (a) The first single-molecule microscope with the optics outside the cryostat; (b) modified commercially available microscope objective immersed in liquid helium; (c) mirror objective especially designed for low temperature single molecule microscopy. Adapted from Ref. 44.

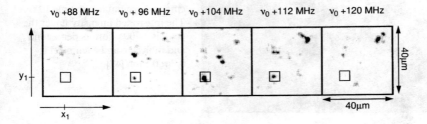

Figure 23. Five spatially resolved histograms, representing an area of $40\,\mu m \times 40\,\mu m$ on the *p*-terphenyl crystal, recorded with the microscope (a) in Fig. 22. Each histogram has been recorded at a different excitation frequency. A small square box of $5\,\mu m \times 5\,\mu m$ size, drawn in each histogram, highlights the rise and decay of the fluorescence intensity of a pentacene molecule as a function of the excitation frequency [41].

three lenses of the microscope objective were mounted into a new holder. The working distance of the microscope was 0.3 mm at a focal length of $f = 2.9$ mm. Between 1.6 and 2.7 K liquid helium has a refractive index of $n = 1.028$, which somewhat modifies the optical beam path. After an initial pre-alignment at room temperature, the image had to be refocused at its operating temperature using a lens outside

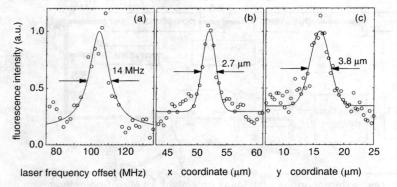

Figure 24. The first graph shows the fluorescence excitation spectra of the pentacene molecule obtained from integration of the square box in Fig. 23. The data were fitted with a Lorentzian function and yield the center of the transition and a width of 14 MHz. The last two graphs present the geometrical intensity distribution of the fluorescence of the same molecule recorded with the microscope (a) in Fig. 22. The Gaussian fit determines the peak intensity positions with a uncertainty of $\pm 0.5\,\mu m$ [41].

Figure 25. Fluorescence microscopy of a single molecule of terrylene in hexadecane. The emission from a round sample of a diameter of $100\,\mu m$ is shown with a fixed laser excitation wavelength of 572.379 nm with intensity 20 mW/ cm^2 and accumulation time 10 s. The integrated intensity under the peak is 4.700 photons per second and molecule and each pixel is $0.7 \times 0.5\,\mu m$ [47].

the cryostat (also changing the magnification). The setup is illustrated in Fig. 22, inset (b).

The first measurements using this microscope were performed on a new system for single-molecule spectroscopy: terrylene in hexadecane [45, 47]. Compared to pentacene in p-terphenyl the system has two major advantages: first the measured maximum emission rate of terrylene is 5 times larger and second, the obtained saturation intensity is 2 orders of magnitude higher (without taking into account the orientation of the molecules [46]), which allows to pump the molecule at higher excitation intensities. Fig. 25 shows a fluorescence microscopy image of a $100\,\mu m \times 100\,\mu m$ area, whereby the z-axis is the fluorescence signal. One terrylene molecule, in resonance with the laser, appears as a very strong peak on a moderate background originating from the host matrix scattering and from terrylene molecules not located in the focal plane of the microscope.

Both improvements, the optimized fluorescence collection and the higher fluorescence yield, led to a significantly higher photon flux. At a wavelength of $\lambda = 610\,\text{nm}$ and a numerical aperture of $NA = 0.85$ the diffraction limited spatial resolution

$$\delta = \frac{0.61 \cdot \lambda}{NA} \tag{11}$$

evaluates to $0.438\,\mu\text{m}$. This value represents the theoretical limit, which the actual experimental resolution may reach. Single molecular emitters are ideal point sources. The width of the spatial intensity distribution of their image on the detector may be interpreted as the 'true' resolution of the imaging system. Presently a best value of $1.6\,\mu\text{m}$ has been achieved for a magnification of 67. Most experiments are performed with a magnification of 40 to 60 giving a resolution of 2.6 to $1.9\,\mu\text{m}$. This experimental resolution is below the theoretical value. The reasons are attributed to the following factors: First, the fluorescence has to pass three cryostat windows behind the microscope objective. Second, an achromatic lens was used to refocus the light onto to the detector. This additional optical element introduces new imaging errors, which modify the optimum working distance between microscope and object. Third, geometrical misalignment of the optical setup leads to a slight asymmetry of the image area of highest resolution. And fourth, the camera itself limits the spatial resolution. Additional test measurements with objects of known size showed, that the horizontal and vertical resolution of the camera was $35\,\mu\text{m}$ and $40\,\mu\text{m}$. At a magnification of 60, the camera alone limits the spatial resolution of the system to about $0.6\,\mu\text{m}$.

The immersion micro objective

A special immersion micro objective was developed for operation under liquid helium conditions [43]. It is a homocentric mirror objective and consists of two suprasil components (see Fig. 22, inset c): The smaller element was fabricated from a full sphere. One of its planar surfaces was polished and served as specimen holder. The distance between this plane and the center of the sphere determines the magnification of the objective. On the other side a second planar surface produced a separation from the reflective coating, which was deposited on the second element. This component forms a homocentric meniscus shell with a convex surface and a small planar central output area. The convex and concave surface were mirror-coated. Both elements are held together by adhesion.

This setup has the following advantages: (i) It is insensitive to vibrations, which occur in the cryostat; (ii) the full light path passes through suprasil and is not affected by immersion in liquid helium; (iii) Suprasil has a small expansion coefficient, which does not change the optical properties of the objective at low and high temperatures and (iv) the shrinking of the optical components leaves the ratios of all dimensions invariant and does not alter its functionality.

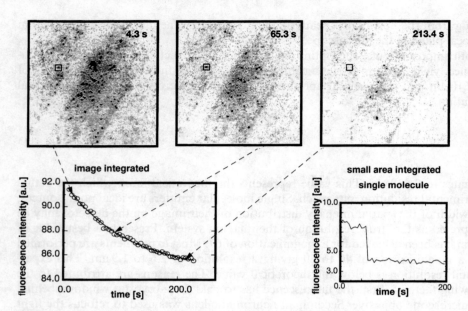

Figure 26. Light induced spectral shifts (equivalent of hole burning in the bulk case) of terrylene molecules in *n*-octane at 1.8 K. An area of approximately $120 \times 120 \ \mu m^2$ is illuminated by ≈ 20.5 W/cm^2 at a wavelength of 571.967 nm. During 220 s of illumination images are recorded using the three lens microscope. One image is integrated over 0.5 s (16 frames), the processing time between the images is at present 3.8 s. Three of them are shown in the upper part. The emitting molecules correspond to dark spots. Their number decreases with the irradiation time. By integrating the fluorescence signal over the whole picture, and plotting the obtained signal at the corresponding times, for all the recorded images, it is possible to extract the kinetics of the photoreaction (lower left figure). The fate of every single molecule can be monitored by integrating a small area of the image corresponding to a molecule. The lower right figure shows such a time trace for the molecule marked with a square in the images. Adapted from Refs. 48 and 49.

Assuming that the sample has the same refraction index as suprasil (about 1.455), the objective has a magnification of $\beta \approx 100$ in lateral and $\alpha \approx 8000$ in axial direction. The numerical aperture is NA = 0.722, but a central part with an aperture of NA = 0.329 is lost due to the Cassegrain mirror design. Its theoretical resolution is 0.52 μm. Small spherical and chromatic aberrations reduce the resolution to 1.0 μm.

1.3.6.2 Applied fluorescence microscopy

Parallel study of the dynamics of single molecules

Fluorescence microscopy allows one to collect the fluorescence from up to 100 single molecules simultaneously. Furthermore it is easily possible to bridge the single molecule regime to the bulk sample regime by integrating the measured property over a large number of fluorescing molecules. Fig. 26 illustrates this method when applied to the "hole-burning" of terrylene in *n*-octane [48, 49]. This system shows a more

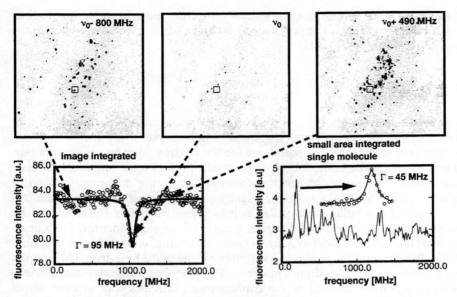

Figure 27. After irradiation, as described in the previous figure, a frequency scan is recorded by taking an image at every frequency point. The upper images are taken at different frequencies, whereby the central one corresponds to the irradiation frequency (lower number of molecules). By integrating the fluorescence signal over the whole picture, and plotting the obtained signal at the corresponding time for all the recorded images, it is possible to extract the hole shape of the burned molecules. By integrating a small area on the images as a function of frequency, the lineshapes of the single molecules present in that area can be obtained as shown in the lower right scan. Adapted from Refs. 48 and 49.

efficient photochemistry than terrylene in hexadecane previously studied. Most of the light induced spectral shifts are irreversible, probably due to the large mismatch between the size of guest and host molecule. The area of the sample ($120 \times 120\,\mu m^2$) was constantly illuminated with approximately $0.5\,W/cm^2$, while recording a time sequence of xy-frames. The upper images in Fig. 26 are a selection taken at different times. It can be seen that the total number of molecules (dark areas correspond to high fluorescence) decreases with time. It is now possible to integrate the signal over the whole image and plot it as a function of time. This is shown in the lower left figure and corresponds, in a certain sense, to a bulk experiment. The fate of every single molecule can be tracked down by integrating a small square in the image, which contains the molecule only, and plotting the results as a function of time (lower right figure). From such diagrams, it is possible to characterize the photostability of every molecule.

Fig. 27 shows three images at frequency $\nu_0 - 800\,MHz$, ν_0 and $\nu_0 + 490\,MHz$, after a 220 s irradiation at ν_0 (see Fig. 26). The central image, taken at the burning frequency, has fewer molecules than the others. By integrating every single image and plotting it as a function of the laser frequency, one can easily access the hole shape (lower left figure). Again the integration over a small square gives a "standard" single-molecule spectrum of a spacial resolution element. The double line-width of a typical molecule is approximately 90 MHz which compares well with the observed "hole width" of 95 MHz.

1.3.7 Outlook

Single-molecule spectroscopy in low-temperature doped solids is a young and rapidly growing field. The experiments presented in this chapter demonstrate the flexibility of this new spectroscopic method and are the starting point to a myriad of new challenging experiments. In the following we briefly present some ongoing studies.

In the attempt to control the spectral dynamics occurring in the millisecond to minute time scale in Shpol'skii systems, recently a new approach has been taken. An attempt was made to build a molecular switch whose frequency shifts, caused by an internal perturbation, are under partial control of the experimentor. Bach et al. [48, 50] investigated the model system terrylene in octane, codoped with triphenylene. A UV lamp excites the perdeuterated triphenylene in the long-lived (20 s) triplet state, causing a change in its dipole moment. The terrylene molecules in the neighborhood are perturbed as long as the triphenylene molecules stay in their triplet state. A rich and puzzling behavior of the single-molecule dynamics was observed.

Still more recently, single diphenyl octatetraene molecules doped into tetradecane were studied, using the frequency-doubled Ti–sapphire laser setup developed [51]. The observed $S_1 \leftarrow S_0$ transition in this system is one-photon-forbidden, but two-photon-allowed: using $1 \, MW/cm^2$ of red light, single diphenyl octatetraene molecules were detected by two-photon absorption. As long as the emission saturation limit is high, single-molecule spectroscopy is still feasible even for low-oscillator-strength molecules.

References

[1] F. Güttler, J. Sepiol, T. Plakhotnik, A. Mitterdorfer, A. Renn, and U. P. Wild, *J. Lumin.* **56**, 29–38, (1993).

[2] F. Güttler, Spektroskopie einzelner Moleküle, Dissertation ETH Nr. 10707, Zürich (1994).

[3] F. Güttler, M. Croci, A. Renn, and U. P. Wild, accepted for publication in *Chem. Phys.*

[4] J. L. Baudour, *Acta Cryst.* **B47**, 935–949 (1991) and references therein.

[5] J. L. Baudour, Y. Delugard, and H. Cailleau, *Acta Cryst.* **B32**, 150–154, (1976).

[6] F. G. Patterson, H. W. H. Lee, W. L. Wilson, and M. D. Fayer, *Chem. Phys.* **84**, 51, (1984).

[7] R. W. Amberger and A. Flammersfeld, *Z. Naturforsch.* **23a**, 311, (1967).

[8] I. Rebane, *Opt Commun.* **110**, 565–568, (1994).

[9] J. Köhler, A. C. J. Brouwer, E. J. J. Groenen, and J. Schmidt, *Chem. Phys. Lett.* **250**, 137, (1996).

[10] M. Pirotta, F. Güttler, H. Gygax, A. Renn, and U. P. Wild, *Chem. Phys. Lett.* **208**, 379, (1993).

[11] W. E. Moerner and W. P. Ambrose, *Phys. Rev. Lett.* **66**, 1376, (1991).

[12] U. P. Wild, A. R. Holzwarth, and H. P. Good, *Rev. Sci. Instrum.* **48**, 1621, (1977).

[13] T. J. Aartsma, J. Morsink, and D. A. Wiersma, *Chem. Phys. Lett.* **47**, 425, (1977).

[14] T. E. Orlowski and A. H. Zewail, *J. Chem. Phys.* **70**, 1390, (1979).

[15] M. Orrit, J. Bernard, A. Zumbusch, and R. I. Personov, *Chem. Phys. Lett.* **196**, 595, (1992).

[16] M. Orrit, J. Bernard, and D. Möbius, *Chem. Phys. Lett.* **156**, 233, (1989).

[17] A. J. Meixner, A. Renn, S. E. Bucher, and U. P. Wild, *J. Phys. Chem.* **90**, 6777, (1986).
[18] M. Pirotta, A. Renn, M. H. V. Werts, and U. P. Wild, *Chem. Phys. Lett.* **250**, 576, (1996).
[19] M. Pirotta, M. H. V. Werts, A. Renn, and U. P. Wild, Helv. Phys. Acta **69**, 7 (1996).
[20] U. P. Wild, F. Güttler, M. Pirotta, and A. Renn, *Chem. Phys. Lett.* **193**, 451, (1992).
[21] J. H. Meyling and D. A. Wiersma, *Chem. Phys. Lett.* **20**, 383, (1973).
[22] J. H. Meyling, W. H. Hesselink, and D. A. Wiersma, *Chem. Phys.* **17**, 353, (1976).
[23] L. Kador, D. E. Horne, and W. E. Moerner, *J. Phys. Chem.* **94**, 1237, (1990).
[24] U. Pürig, D. W. Pohl, and F. Rohner, *J. Appl. Phys.* **59**, 3318, (1986).
[25] E. Betzig, J. K. Trautman, T. D. Harris, J. S. Weiner, and R. L. Kostelak, *Science* **251**, 1468, (1991).
[26] E. Betzig and R. J. Chichester, *Science* **262**, 1442, (1993).
[27] J. K. Trautman, J. J. Macklin, L. E. Brus, and E. Betzig, *Nature* **369**, 40, (1994).
[28] W. E. Moerner, T. Plakhotnik, T. Irngartinger, U. P. Wild, D. Pohl, and B. Hecht *Phys. Rev. Lett.* **73**, 2764, (1994).
[29] T. V. Plakhotnik, *Optics and Spectroscopy* **79**, 688, (1995).
[30] A. P. Fein, J. R. Kirtley, and R. M. Feenstra, *Rev. Sci. Instr.* **58**, 1806, (1987).
[31] H. W. Offen in *Organic Molecular Photophvsics* (Ed.: J. B. Birks), Wiley, London, 1973, p. 103 and ref. therein.
[32] B. J. Baer and E. L. Chronister, *J. Chem. Phys.* **99**, 3137–3138, (1993).
[33] K. Holliday and U. P. Wild, in *Molecular Luminescence Spectroscopy Methods and Applications: Part 3* (ed. S. G. Schulman), Wiley 1993, p. 149.
[34] T. Sesselman, W. Richter, and D. Haarer, *Europhys. Lett.* **2**, 947, (1986).
[35] M. Croci, H. J. Müschenborn, F. Güttler, A. Renn, and U. P. Wild, *Chem. Phys. Lett.* **212**, 71, (1993).
[36] S. Kummer, Th. Basché, and C. Bräuchle, *Chem. Phys. Lett.* **229**, 309, (1994).
[37] A. Müller, W. Richter, and L. Kador, *Chem. Phys. Lett.* **241**, 547, (1995).
[38] A. Amirav, U. Even, and J. Jortner, *Opt. Comm.* **32**, 266, (1980).
[39] W. P. Ambrose and W. E. Moerner, *Nature* **349**, 225, (1991).
[40] W. P. Ambrose, Th. Basché, and W. E. Moerner, *J. Chem. Phys.* **95**, 7150, (1991).
[41] F. Güttler, T. Irngartinger, T. Plakhotnik, A. Renn, and U. P. Wild, *Chem. Phys. Lett.* **217**, 292, (1994).
[42] T. Irngartinger, A. Renn, and U. P. Wild, J. Lumin. **66 & 67**, 232, (1996).
[43] J. Jasny, J. Sepiol, T. Irngartinger, M. Traber, A. Renn, and U. P. Wild, Rev. Sci. Instrum. **67**, 1425, (1996).
[44] M. Croci, T. Irngartinger, A. Renn, and U. P. Wild, *Exp. Techn. Phys.* **41**, 249, (1995).
[45] T. Plakhotnik, W. E. Moerner, T. Irngartinger, and U. P. Wild, *Chimia* **48**, 31, (1994).
[46] T. Plakhotnik, W. E. Moerner, V. Palm, and U. P. Wild, *Opt. Commun.* **114**, 83, (1995).
[47] W. E. Moerner, T. Plakhotnik, T. Irngartinger, M. Croci, V. Palm, and U. P. Wild, *J. Phys. Chem.* **98**, 7382, (1994).
[48] H. Bach, oral presentation at the Workshop: "Single Molecule Spectroscopy: New System and Methods" held in Ascona (Switzerland), 10–15 March 1996.
[49] H. Bach, T. Irngartinger, A. Renn, and U. P. Wild in preparation.
[50] H. Bach, T. Irngartinger, A. Renn, and U. P. Wild in preparation.
[51] T. Plakhotnik, D. Walser, M. Pirotta, A. Renn, and U. P. Wild, *Science* **271**, 1703, (1996).

1.4 Spectral Jumps of Single Molecules

R. Brown, M. Orrit

1.4.1 Introduction

Among the most striking points about single-molecule lines are the sudden intensity changes they undergo. These jumps in fluorescence intensity, which in fact arise from jumps of the molecular frequency with respect to the exciting laser, are intrinsically irreproducible and random and have no equivalent in conventional spectroscopy of large populations of molecules in bulk systems. Spectral jumps provide one of the most convincing arguments for the attribution of narrow fluorescence excitation structures to single molecules (see Sections 1.1 and 1.2). Spectral jumps are very direct indicators of the nanoscopic dynamics in the area surrounding the single molecule probe under study, each single molecule probing a different area. In this Section, we review the experimental results that were obtained in the last few years about spectral jumps and the main conclusions drawn from them. The next Section (1.5) by J. L. Skinner, will describe the theoretical approach to these dynamics problems, and the new theoretical methods that were specifically devised to deal with single molecules.

1.4.1.1 Sensitivity of zero-phonon lines to the environment

One of the main values of single molecule studies is the possibility of focusing on the microscopic environment of the molecule up to a radius of some tens of nanometers. By isolating only one of these regions in a large sample, information is obtained in a very direct way, which avoids all the averaging entailed by classical methods.

Here, we consider molecules in solids at low temperatures. For these systems, the existence of narrow zero-phonon lines [1] has two important consequences. First, the absorption cross-section is enhanced by several orders of magnitude at resonance, aiding the detection. Second, since the zero-phonon line is very narrow, it is very sensitive to external perturbations and to changes in the molecular environment. Perturbations may affect the intensity and the width of a single molecule line, but the most sensitive feature of the line is its position: frequency shifts corresponding to a molecule's natural width correspond to less than a tenth of a millionth change in its transition frequency. Such tiny shifts can be measured with the high accuracy of a single-mode laser. If the shifts of the resonance frequency are larger than or comparable to the molecular linewidth (typically a few tens of MHz) and if these changes occur on a long enough time scale, they lead to easily measurable changes in the spectrum, corresponding to very subtle changes in the molecular environment.

Let us first estimate the order of magnitude of the interactions of the molecule with its environment, leading to measurable spectral changes. As discussed in Section 1.3, frequencies of single molecule lines are sensitive to an applied static electric field [2, 3]. Even centro-symmetric molecules, when included in a disordered matrix where their symmetry is broken, can acquire a dipole moment change of about 1 D and thus shift at a rate of about 1 MHz for a field of 1 kV/m [3]. Such fields are extremely small by molecular standards. For example, an electron or a proton moving by 0.1 nm at a distance of 10 nm will shift the single molecule line by 50 MHz, enough to "feel" the motion as a spectral change. Alternatively, a weakly polar chemical bond such as a C-H (dipole moment of 0.1 D), rotating through a small angle of, say, 10° is still "felt" 1.5 nm away by the molecule, i.e. beyond its first solvent shell. Similarly, pressure effects are also felt with high sensitivity. The pressure shift is of the order of 1 GHz per atmosphere [4], which means that a pressure change of $5 \cdot 10^3$ Pa will shift the resonance by 50 MHz. Again, this is a very weak stress on the molecular scale. Let us assume a deformation or strain of 0.01 nm brought about by some change in a small region of 0.3 nm radius. Local elastic perturbations decrease like the inverse cubic power of distance [5, 6]. With an elastic modulus of 10^{10} Pa, typical of molecular solids, the localized change still induces a stress of $5 \cdot 10^3$ Pa some 12 nm away from its origin, enough to shift or broaden a molecular line. Glasses present specific low-temperature excitations known as two-level systems (TLS). Flipping one of these systems with deformation potential D induces a stress of approximately Dr^{-3} at distance r. For a deformation potential of 0.3 eV, a typical value in organic glasses [7], the stress shifts the molecular line by 50 MHz at a distance of 24 nm. These examples show that single molecules will be sensitive to motion at relatively large distances. If the dynamical processes perturbing the molecule are slow enough (slower than the dephasing time T_2), the resonance frequency can be considered as a classical function of time. Thus, it can be used to probe the dynamics of the environment. The picture will be clearest when most of the surroundings are frozen, and just one localized area is moving. This will be the case at low temperature, if we consider the motion of light particles, like electrons and protons [8], or the coordinated motion of several nuclei responsible for TLS dynamics in glasses.

1.4.1.2 Dynamics of disordered solids: two-level systems

The dynamics of liquids or of solids at room temperature involve many degrees of freedom, some with periods as short as tens of femtoseconds, others with much longer characteristic times. Spectroscopic investigations at room temperature nowadays use very short light pulses, and their analysis must include the many active modes in statistical ways. The experimental results we shall review here, being restricted to the dynamics of solids at low temperatures, are situated at the other end of the spectrum of experimental performance. Their time resolution will be rather modest (from microseconds to seconds), but their spectral resolution is very high. Although the systems studied are still very complex, and present many degrees of freedom, most of them are frozen, leaving but a few active ones to be studied in greater detail.

Dynamics depend crucially on structure. Crystals having – by definition – well-defined structures, they usually leave little room for low-energy motion, apart from acoustical phonons. Because acoustical phonons have long wavelengths and a low density of states, their direct effect on localized impurities often are practically negligible. There are a few exceptional crystals in which significant motion still takes place at liquid helium temperatures; for instance, when loose and light atomic groups can tunnel between different configurations in the bulk of a crystal. This is the case for methyl groups, which jump between different magnetic-rotational states in their threefold potential well [8–10]. In most cases, however, low temperature dynamics will arise from defects, either those occuring in real, disordered crystals, or those present among the very complex configurations of the highly disordered, amorphous solids known as glasses. While the subject of special crystal defects must be dealt with in each specific case (see the next section and Section 1.5), the distinctive features of glass dynamics appear to be common to many systems. We want now to briefly review what those features are and how they were identified.

It became clear some 25 years ago that many low-temperature properties of amorphous solids (glasses) differ fundamentally from those of their crystalline counterparts [11]. In high-quality crystals, a Debye model of acoustical phonon branches accounts very well for observed properties such as the specific heat. Glasses as solids also possess acoustical phonon branches. Since long wavelength acoustical phonons are not sensitive to the short-range disposition of atoms, which was thought to be the main difference between crystals and glasses, the low-temperature properties of glasses were expected to be quite similar to those of crystals. Instead, many properties of glasses were found to present strong anomalies (e.g. their specific heat and conductivity), showing that glasses present extra degrees of freedom, which have no equivalent in crystals. Glasses, as opposed to crystals, do not possess a single, well defined structure. The potential energy surface of an ensemble of N particles is extremely complicated, with a number of local minima growing exponentially with N [12], of which the lowest and most symmetrical ones are crystalline structures. When a glass-forming liquid is quenched through its glass transition temperature, it is kinetically trapped in narrower and narrower regions of the energy surface, down to the scale of a single minimum at very low temperature. Transitions between these minima give rise to rich dynamics, even at low temperature, because classical activation over potential barriers is replaced by quantum-mechanical tunneling through the barriers.

The excess specific heat of glasses varies linearly with temperature, whereas the Debye contribution varies as T^3 and is therefore negligible at low temperature. This indicates that the density of states of the extra degrees of freedom is roughly constant for low energy in the range of up to a few Kelvin. Moreover, since the absorption of ultrasound waves or of microwaves by glasses saturates at high intensity [11], one has to conclude that these excitations behave like strongly anharmonic oscillators or like two-level systems. These observations are among the main arguments for the model of the two-level systems (TLS) proposed independently in 1972 by Phillips [13] and Anderson et al. [14] to describe glass anomalies.

Some years later, high-resolution spectroscopy of solids started to develop quickly after the advent of narrow and tunable dye lasers. New techniques such as photon echoes [15, 16] and hole-burning [17–19] were applied to probe molecules in glassy systems. In addition to a strong inhomogeneous broadening, reflecting the wide variety of local environments and the lack of any order, glasses also exhibit very broad homogeneous widths, even at low temperatures [20]. The study of optical properties of impurities in glasses is more complicated than that of the pure glasses, because of the perturbation of the matrix by the optical probe. For example, significant changes of shape of the probe may occur upon excitation, and a large amount of energy is released in non-radiative transitions. These effects lead to non-photochemical hole burning processes [21], in which the dye–matrix system jumps to a new ground state configuration, different from the initial one. Despite these new features, it soon appeared that the large homogeneous widths and their weak temperature dependences were closely related to the anomalies found in the many other physical properties of glasses [22, 23]. In early optical experiments, no intermediate timescale was experimentally available between the picoseconds of photon echoes and the many seconds of persistent spectral hole burning. Dephasing was naturally stressed as the main broadening cause, as is the case for molecular crystals above a few K [24]. More recent experiments, however, have shown that the timescales of optical broadening in glasses are spread over many decades, and that, in addition to dephasing, spectral diffusion plays a dominant part in many hole burning and photon echoes experiments [25–27].

The additional degrees of freedom of glasses, which play the dominant part in the low-temperature dynamics, are related to transitions between local wells of the potential. These wells are thought to be separated by energy barriers of the order of the glass transition temperature [13]. The barriers can be crossed by classical activation at high temperature, but at liquid helium temperatures the thermal energy is at least one order of magnitude less than typical barrier heights, so they can be crossed only by quantum-mechanical tunneling. The attribution of glass dynamics to tunneling at once explains the wide spread of relaxation times observed in ultrasound absorption and dielectric relaxation, as well as in optical experiments. Indeed tunneling probabilities depend exponentially on barrier characteristics (height V and width d along the generalized coordinate) and on the mass m of the tunneling particle as:

$$p \propto \exp(-2d(2mV)^{1/2}/\hbar)$$

Since barrier parameters in a disordered system are broadly distributed, tunneling rates are spread over many orders of magnitude, from picoseconds to hours or longer.

The most probable tunneling event from a given instantaneous configuration of the glass should thus involve small movements of a few atoms, i.e. they should be local processes. Since a tunneling event changes the strain in the neighborhood, it may open new tunneling pathways for some atoms or modify or deny tunneling pathways for others. Thus, TLS's should not in general be independent. However, at sufficiently low temperatures, the number of active tunneling systems should be small enough to neglect their interactions. The average separation of tunneling systems active at 1 K, deduced from the specific heats of a variety of glasses [11], is about 150 Å. We thus obtain a model of the glass as a set of (weakly interacting) TLS's, which tunnel independently between configurations separated in energy by a few $k_B T$ at most [13, 14]. For sufficiently low temperatures, the concentration of tunneling systems will be small, and the model of independent TLS will describe the glass satisfactorily [11]. At higher temperatures, the concentration of flipping TLS will be too large to be consistent with the absence of interaction between them. In this regime, hierarchical interactions between regions may set in [28], ultimately leading to the glass transition.

Tunneling being very sensitive to barrier characteristics, it might seem that low-temperature dynamics should depend very strongly on the chemical nature of the glass. Polymers, for example, where movements should be coordinated at long distances by the long backbones, might behave quite differently from glasses of small molecules, where only short-range interactions are present. However, experiments show that the dynamical characteristics are very similar in a variety of amorphous systems, with quite different chemical natures. The spatial and energetic density of tunneling systems is comparable in many glasses, including fused silica, metallic alloys, polymers, etc. Molecular dynamics simulations of an atomic alloy glass with Lennard Jones interactions yield a density of TLS's which is compatible with experiments [29]. The insensitivity of low-temperature dynamics to the details of the chemical nature and the interactions suggests that they are determined by geometrical properties of the arrangement of atoms or groups of atoms in solids with quenched disorder. The nature of this arrangement, or the kind of generalized coordinate describing tunneling are still open questions, after more than two decades of glass studies.

In order to model glass properties and to discuss experimental observations, we need a model of tunneling systems. A schema of a TLS, and its potential along the generalized coordinate is presented in Fig. 1. Independently of the nature of this coordinate, which determines the mass m of the tunneling particle, the TLS is characterized by the barrier width d (taken as the distance between minima in each well), the barrier height V, and the asymmetry Δ or energy difference between the wells,

Figure 1. Model of a probe molecule surrounded by tunneling systems. The microscopic parameters such as the barrier height, V, the width, d, the mass of the tunneling atoms m, are distributed quantities in a glassy system, determining a wide spread of the tunneling splitting ΔE and the tunneling matrix element between the localized states in either well.

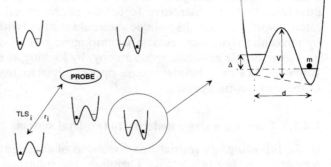

supposed to be much less than V. Restricting ourselves to the lowest energy states in the potential of Fig. 1, tunneling is essentially described by the tunneling matrix element Δ_0, which decreases exponentially with d, and with $(mV)^{1/2}$. The tunneling systems of a real glass are determined by the actual disordered configuration, and therefore have a broad range of asymmetries. Although working at a few Kelvin restricts the set of TLS's to those with asymmetries comparable to $k_B T$, the asymmetry will usually be much larger than the tunneling matrix element Δ_0. In this case, coherent tunneling only leads to weak mixing of the localized states in either well. If acoustical phonons are present, the asymmetry and tunneling element of the TLS fluctuate and transitions take place between the quasi-localized states. These incoherent tunneling processes, assisted by acoustical phonons, will predominate tunneling from one well to the other at liquid helium temperatures. Different tunneling processes are obtained by expanding the TLS parameters as functions of phonon coordinates. The change of asymmetry Δ under pressure variations leads one to define a deformation potential, which determines the interactions of the TLS with the phonon bath [30]. In a similar way, the dipole moment change between the two wells describes the interaction of the TLS with microwaves responsible for dielectric relaxation, microwave photon echoes, etc. [31]. The lowest order in perturbation theory of the TLS-phonon interaction gives the direct process, in which the TLS jumps from one well to the other by absorbing or emitting a single phonon. To second order, the TLS may jump by jointly absorbing and emitting two different phonons, in so-called Raman processes. At high temperature, the TLS may interact with many phonons in a strong coupling regime [32, 33]. The different processes have distinct temperature dependences, which can help distinguish them in actual experiments.

The standard TLS model assumes that Δ and $\log(\Delta_0)$ are distributed evenly and are uncorrelated, and that the systems are independent and randomly distributed in space. With adjusted distributions, it describes the anomalous glass properties correctly, though phenomenologically. Yet, the standard TLS model leaves many questions open, such as the mechanisms of interaction between TLS's or between a TLS and phonons, or such as the dynamical equilibrium within the distribution of TLS's in a macroscopic glass sample. The most fundamental and difficult problem is that of the nature of TLS motion, and its seemingly general appearance in chemically different glasses. All these questions are difficult to answer by means of conventional measurements because they involve large numbers of TLS's, yielding only average quantities that are insensitive to details of the model. The combination of more microscopic methods, like single molecule studies, with the simulations of glasses by molecular dynamics seems a promising way of shedding light on these difficult problems. Single molecule spectroscopy, by looking at a small area around an individual local probe, provides unique opportunities to investigate more specific details of the microscopic systems.

1.4.1.3 Coupling a single molecule to two-level systems

In the following, we restrict our discussion of single molecule spectra to the lowest temperatures, say below 5 K. Therefore, the influence of matrix dynamics on the probe molecule can be described in terms of TLS's. Let us first consider the inter-action of the two electronic states of the molecule with a single TLS. In general, we would have a four-level system, with different rates of possible optical transitions between the lower two and upper two levels, and with relaxation between levels. Such a scheme may be treated exactly by a Green's function formalism [34]. How-ever, the roles of molecule and TLS's are highly asymmetrical: When the TLS flips, the molecular resonance with the laser is dramatically affected, whereas the flipping of the TLS due to interaction with phonons is hardly affected by the molecule being excited (we disregard photo-induced jumps in the present discussion). It is therefore warranted (for spontaneous jumps) to neglect the influence of the molecule on the TLS, i.e. to assume that the electronic system plays the role of a probe of the TLS. Furthermore, the principal effect of the nuclear TLS is to shift the molecular reso-nance, so that its frequency can be followed as a semi-classical function of time, reflecting the state of the TLS.

When a TLS jumps incoherently from one well to the other, the duration of the jump is usually much less than the dwell time in the wells. The duration of barrier crossing is of the order of the oscillation period in one of the wells, while the time between jumps is much longer. We may then use the Kubo-Anderson [35, 36] model of the lineshape, where the molecular frequency is modulated by sudden jumps between two values separated by $\delta\omega$, and where τ_c is the average delay between jumps. In the case $\delta\omega\,\tau_c \ll 1$, the time between jumps is so short that individual jumps can only slightly dephase the electronic oscillation. In this fast modulation limit, one obtains a single line at the average frequency, with a width proportional to the rate of phase loss due to the randomness of individual jumps, i.e. inversely pro-portional to the jump rate. This phenomenon is at the root of dephasing, common in molecular crystals, where fast modulation occurs due to localized phonons (libra-tions). In the opposite limit of slow jumping, $\delta\omega\,\tau_c \gg 1$, the molecular frequency successively assumes its two distinct values long enough to be measured during the dwell time between jumps. This leads to two lines in the absorption spectrum, the molecular absorption line jumping from one position to the other with the driving TLS. The line diffuses in the spectrum, hence the name spectral diffusion given to this limit of slow modulation. Spectral diffusion has now been observed in a number of systems, in particular glassy ones where relaxation proceeds over a very broad range of time scales. The modulation is classified as fast or slow depending on the magnitude of the frequency jump, i.e. on the distance of the driving TLS to the mol-ecule. If we only discuss jumps large enough to be observed directly, with $\delta\omega > \gamma_h$ (γ_h is the homogeneous width), we see that modulation will always be slow for $\tau_c > T_2$, of the order of nanoseconds or shorter [26]. The spectral jumps observed on single molecules so far had amplitudes in excess of 100 MHz and characteristic times in the range from microseconds to hours, well within the slow modulation domain.

In a real disordered matrix, there usually are several TLS's able to interact with a single molecule, as schematically shown in Fig. 1. The first case is that of defects in crystals, where a non-random distribution of flipping TLS's in space interacts with the probe. Dislocations or grain boundaries can give rise to such one- or two-dimensional distributions of TLS's (see Section 1.4.2). Because of the translational symmetry of crystals, we may assume in this case that all TLS's have the same parameters (asymmetry, rates of tunneling, etc.). The distribution of frequency jumps will reflect that of TLS's in the molecular neighbourhood, with a maximum jump size related to the distance of the probe to the domain wall, while a single characteristic time will appear, for example in the frequency auto-correlation function. On the other hand, if the probe is included in a glass, it interacts with TLS's which are not only distributed randomly in space (we may at first assume this in the absence of any detailed model for glass structure and for the nature of TLS's), but also have distributed parameters. Jumps with a broad range of magnitudes and of timescales are then expected to appear in the spectral diffusion of the probe. Moreover, spectral diffusion is expected to vary considerably when going from one molecule to another, depending on the particular distribution of TLS's around the probe.

Let us now consider the case where jumps of a nearby TLS are driven by excitation of the probe molecule. From the experimentalist's point of view, if the back reaction occurs over times longer than minutes, the reaction will be irreversible. It is extremely difficult in this case to distinguish between spontaneous and photo-induced jumps, since both processes may coexist, and each molecule can be probed only once. However, the measurement of tens or hundreds of single molecules with different exciting intensities usually shows without ambiguity if there are photo-induced jumps or not. In bulk systems, these jumps give rise to persistent spectral hole-burning. In a hole-burning experiment, a large population of molecules is excited at a given wavelength in an inhomogeneous band, leading to a depletion of absorbing molecules at the burning wavelength, i.e. to a spectral hole. Since the effect of spontaneous jumps is compensated in this case by the dynamical equilibrium of the inhomogeneous distributions, only the photo-induced jumps contribute to the spectral hole. Several mechanisms lead to persistent hole-burning [21, 37, 38]. In photochemical hole-burning, excitation of the probe may lead to breaking or rearrangements of chemical bonds either within the molecule or between molecule and matrix, with dramatic frequency changes. Such mechanisms are practically irreversible at liquid helium temperatures. A second kind of burning mechanism is non-photochemical (or photophysical), in which excitation leads to subtle changes of configuration in the molecule–matrix system [38]. This kind of mechanism has been postulated and confirmed for solutions of stable molecules in amorphous matrices, where flipping of nearby TLS's can be triggered by optical excitation. Non-photochemical reactions may be reversible. This means that a spectral hole may refill as time goes, with a broad range of timescales reflecting the dispersion in parameters of the activated TLS's [39]. For a single molecule, photo-induced jumps of a nearby TLS, with a sufficiently fast back-reaction, can resemble spontaneous jumps, but their photo-induced character can be checked by varying the excitation intensity. So far, the study of single molecules has focused on chemically stable molecules, for which only photophysical burning could occur.

1.4.1.4 Experimental methods for spectral jump studies

Once a single molecule line has been isolated with sufficient signal/noise ratio in a fluorescence excitation spectrum (Chapter 1.1), it may or may not exhibit instability due to spectral jumps [40]. Large irreversible jumps move the molecular line outside the accessible range of frequencies. Such one-shot measurements are difficult to exploit; they must be repeated on a large number of molecules, for example in order to determine if the jumps are photoinduced or not [41]. Sometimes, jumps with slow back-reactions occur within the scanning range of the laser. Then, jump rates can be studied by recording statistics of direct and back reactions [42]. The most practical case for the study of spectral jumps, however, is when back-and-forth jumps occur on a timescale shorter than a few seconds. First, by looking at their dependence on excitation power, one can determine whether they are photo-induced or not. The study of photo-induced jumps may then proceed as the exciting intensity is varied. For spontaneous jumps, the most direct method is to record the line position as a function of time, by repeatedly scanning the laser [41]. For a sufficient signal/noise ratio, trajectories of the line in the spectrum will be obtained. If the timescale of the jumps is long enough, these trajectories hold all the available information about the interaction between the molecule and TLS's. The trajectories can be used to deduce the auto-correlation function of the frequency, the distribution of jumps, or other quantities which can be compared to theoretical models [43]. However, if jumps are too fast (faster than seconds to milliseconds, depending on the available fluorescence signal and on the background), the line position cannot be accurately determined in a single scan because of noise. It is still possible to probe jump dynamics by forgoing complete information about the line positions: by keeping the laser frequency fixed on the average position of the line, frequency jumps will be translated into jumps of the fluorescence intensity [44]. The time dependence of the fluorescence signal also contains information about the jumps, albeit distorted by the (often Lorentzian) lineshape of the molecular homogeneous line. A convenient way to exploit this information is to measure the intensity auto-correlation function, for which dedicated electronics exist. This function can then be compared to theoretical models (Section 1.5).

1.4.2 Spectral jumps in crystals

1.4.2.1 Crystal structure and defects

Crystals at low temperatures have well-defined, stable structures. The heat capacity of insulating crystals, for instance, follows Debye's law very well, which means that acoustical phonons are the only excitations that matter at liquid helium temperatures. Therefore, it came as a surprise that spectral jumps appeared in a crystal, as soon as single molecule lines became readily observable [40] (see Fig. 2). The perfect crystal structure being an absolute potential minimum, around which motion should be harmonic, jumps between metastable configurations had to be somehow related to crystal defects [41]. Thermodynamics indicates that defects must be present in any crystal, as it must grow at a high temperature for kinetic reasons. Most crystal defects (vacancies, dislocations, grain boundaries, etc.) are frozen at liquid helium temperatures, and only contribute to a static spread of the frequencies of impurity centers in crystal matrices. The presence of defects is vividly revealed by the inhomogeneous broadening of the optical transitions of substitutional impurities. But in

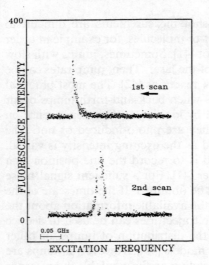

Figure 2. Example of spectral jumps of a single pentacene molecule in *p*-terphenyl, interrupting emission during two scans of the laser (reproduced from Ref. 40).

addition to static broadening by disorder, defects may also have dynamical effects. First, whereas the perfect crystal packing would be too tight for motion to occur, the faulty molecular packing around defects creates room for molecular motion. Second, since the structure of a defect can be very complex, it may easily present configurations with nearly degenerate energies, between which tunneling may take place, even at very low temperatures. A straightforward instance is moving a defect by one lattice step: this does not require energy (in the absence of other defects), but the motion can be felt by a probe molecule in the crystal. Because the structure of perfect crystals is known, and although that of defects is largely unknown, crystals seem in principle more favorable than glasses to understand the microscopic nature of the complex movements responsible for spectral diffusion. In the following, we consider the results obtained with single molecules in the last few years, first in *p*-terphenyl and then in other crystals.

1.4.2.2 Pentacene in *para*-terphenyl crystal

The structure of the *p*-terphenyl crystal at ambient conditions is monoclinic (P21/a) with two symmetrically arranged molecules in the unit cell [45–47]. In each molecule, the rotation of the central phenyl ring around the single bonds connecting them to the outer rings is not free: it is sterically hindered by hydrogen atoms in the ortho position. Therefore, there are two symmetric equilibrium positions of the central ring, tilted by about 13°, determined by intramolecular electron conjugation and intermolecular crystal forces on one hand, and hydrogen repulsion on the other [46]. In the monoclinic phase, the two positions are equivalent, so that the molecule is planar on average (though with an abnormally large libration amplitude [47]). When the crystal is cooled, it undergoes an order-disorder phase transition at 193 K to a new, triclinic phase (P$\bar{1}$) with four inequivalent molecules in the unit cell [45]. The two positions of the central ring are no longer equivalent, which means that the rings are tilted on average by different angles for the four molecules. Tilt angles for neighbors along *a* and *b* have opposite signs (antiferroelastic order). The symmetry breaking occuring during the transition may happen in different ways, leading to four types of energetically equivalent domains [48]. Two of them are related by symmetry in a glide plane, and are distinguishable crystalline structures. The two others are deduced by translation by half the lattice constant. Different kinds of walls must therefore exist between these domains, and can present dynamical properties quite different from the bulk.

Since the size and shape of pentacene and *p*-terphenyl are similar, one expects pentacene molecules to replace one of the four inequivalent terphenyl molecules in the low-temperature crystal. The absorption and fluorescence spectra of the mixed crystal indeed show four electronic origins or sites labelled O_1 to O_4 [49]. The lowest energy sites O_1 and O_2 may be attributed to the less distorted sites on the basis of molecular packing and phonon side-band calculations [50, 51]. The less distorted sites O_1 and O_2 are the interesting ones for single molecule spectroscopy because of their low intersystem crossing (see Section 1.2). The presence of crystal defects appears from the inhomogeneous broadening of these absorption lines. Their band-width is about 40 GHz in melt-grown samples [52], whereas it can be as small as 1 or 2 GHz in sublimation-grown crystals which have been carefully handled [53, 54]. On the other hand, when glued on the end of a fiber and subject to thermal cycles, thin sublimation flakes can present inhomogeneous widths in excess of several cm^{-1} [55].

The inhomogeneous band shape, being static, does not give any hint about the dynamical processes which may affect individual single molecules. Single molecule studies for the first time showed the discrete nature of spectral diffusion, which seldom occurs in crystals. Orrit and Bernard [40] observed sudden spectral jumps as changes in fluorescence intensity for fixed laser frequency. They used the jumps as a further proof that the signals arose from single molecules, ascribing them to re-arrangements in the molecular neighborhood [56], involving the mobile middle ring of *p*-terphenyl. As discussed below, Moerner and colleagues later demonstrated that jumps in the present case are spontaneous [57, 41], i.e. induced by interaction with the phonon bath rather than by the optical excitation of pentacene.

Different patterns of spectral jumps that appear in the pentacene/*p*-terphenyl system were studied in detail by Ambrose, Basché and Moerner [41]. Let us summarize the general appearance of these spectral jumps described in Ref. 41 and later work. Several of the experiments described in Section 1.1–1.3 (e.g. magnetic resonance) require stable molecules, i.e. molecules with transition frequencies which do not appreciably change over several minutes. Fortunately, such stable molecules are common in pentacene-doped *p*-terphenyl crystals, even in those subject to high strain from gluing to an optical fiber. The width of single molecule lines in strained crystals does not always reach the lifetime limit deduced from photon echoes measurements [58] (8 MHz), even after laser drifts have been carefully eliminated [59]. The additional broadening was attributed to spectral diffusion, i.e. even stable molecules feel the dynamics of remote defects, which shift their frequencies. The lifetime limited width could be recovered in high quality crystals [57, 60] (although in this case slow spectral diffusion may have been eliminated together with laser drifts when the single

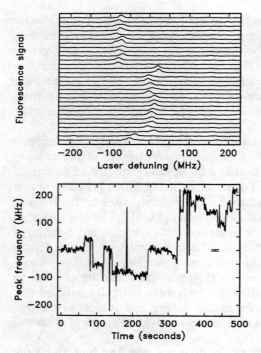

Figure 3. Example of a "jumping" pentacene molecule in p-terphenyl (class II, see text), for which the frequency undergoes irregular jumps. Top: successive excitation scans of the laser; bottom: Molecular resonance frequency as a function of time (reproduced from Ref. 41).

molecule line was swiftly scanned and the maxima of the resulting spectra were superimposed to improve the signal to noise ratio).

When discussing the properties of single molecules, Moerner and coworkers distinguish between stable molecules, which do not perceptibly jump during the few minutes that a measurement lasts (class 1 molecules of [41]), and jumping molecules (class 2). An example of the frequency trajectory of a class 2 molecule is shown in Fig. 3. This distinction is based on a comparison of the spectral range of the jumps with the homogeneous width, i.e. on the distance of the nearest perturbing regions of the crystal. The authors of Ref. 41 found a majority of stable molecules near the center of the inhomogeneous band, while the fraction of class 2 molecules increases as one explores further into the wings. For example, nearly unstrained crystals present some 40% of jumping molecules at 196 GHz to the red from the line center, but only 23% at 55 GHz and none a few GHz away from the center. Stable molecules allow a number of steady state measurements which would be very difficult or impossible on jumping molecules. For instance, the single molecule linewidth increases exponentially at temperatures higher than 4 K, due to interaction with librations. The same effect must exist for jumping molecules, but is very difficult to measure. Stable pentacene molecules can be observed at high intensities for many minutes or hours, which shows that the yield of photo-induced spectral jumping must be extremely low (less than 10^{-9}). In fact, no clear case of photo-induced jumps could yet be documented in this system, despite a rich array of spontaneous jump patterns, to which we turn now.

Figure 4. Example of a "creeping" molecule, showing spectral diffusion toward the center of the inhomogeneous line (from Ref. 41).

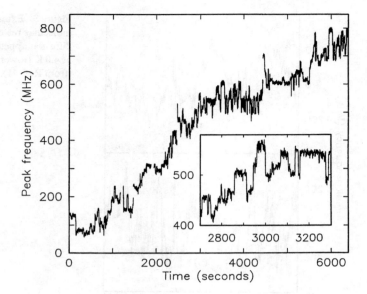

Besides stable lines which reappear scan after scan of the laser, some structures change dramatically between two scans. They can be attributed to spectrally jumping molecules, which become more frequent in the wings of the inhomogeneous distribution, but can be found together with stable molecules in the same sample volume (about $100\,\mu m^3$) and at the same frequency. The occurence of jumping molecules can be coarsely described by postulating two classes of molecule, with different inhomogeneous distributions, the distribution of jumping molecules being very broad. The fluorescence signal of a jumping molecule as a function of time for fixed laser frequency presents discontinuities and looks like a random telegraph signal [60]. The discontinuities show that the frequency changes can be regarded as sudden jumps, at least on the time scale of the experiments, of the order of milliseconds (their appearance is quite similar to that of the quantum jumps displayed by single trapped ions [61]). These findings disagree with a picture of spectral diffusion where molecular frequencies would be slowly modulated by continuous relaxation of the environment, but are consistent with the tunneling system model, with its discontinuous tunneling jumps between different configurations.

By rapidly scanning the laser and determining the resonance frequency of the molecule on each scan, Ambrose, Basché and Moerner obtained trend curves or trajectories of the frequency as a function of time [41], examples of which are shown in Figs. 3, 4 and 5. These trends have a time resolution of about 1 s, but may be accumulated over many minutes or hours to give long-term trajectories. They contain a wealth of information about the distribution of perturbing centers responsible for the jumps in the neighborhood of the molecule studied. Most of this information is lost to conventional measurements because of averaging. For example, averaging over time of the trajectory as in measuring the frequency auto-correlation function, leads to a loss of information compared to the trajectory itself [43]. The average over the

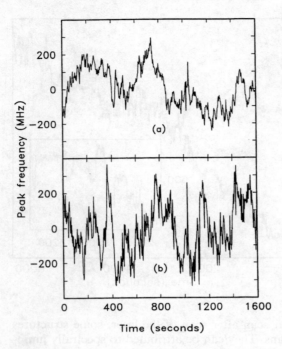

Peak frequency (MHz)

Time (seconds)

Figure 5. Example of a "wandering" molecular resonance frequency. Spectral diffusion appears to be faster and greater at 4.0 K (lower) than at 1.5 K (upper) (from Ref. 41).

many molecules probed in a bulk sample suppresses even more information, for example the discontinuous nature of spectral diffusion. By studying different molecules, the authors of [41] found different behaviors:

A rather common case for unstrained crystals, the "jumper", is that of a molecular resonance performing a random walk within a set of preferred frequencies. In its simplest version, the two-state jumper, the molecular resonance visits only two frequencies, i.e. it is coupled to a single TLS. The two-frequency case is rather rare in pentacene/*p*-terphenyl (one case in a few tens). The more frequent multistate jumpers usually explore a large number of frequencies, which may be explained by a coupling of the molecule to several TLS's.

When the number of perturbing defects in the molecular surroundings is large, either because the molecule is very close to an extended crystal fault, or because the crystal is extremely disordered, the number of preferred frequencies is so large that it becomes impossible to recognize a "preferred" set. Then, the number of jumps between two scans of the laser is large, and the trajectory looks like a one-dimensional random walk, or a spectral diffusion. Such a pattern appears for jumping molecules in strained and disordered crystals at the end of an optical fiber.

Spectral diffusion behaviors differ deeply from molecule to molecule. For example, one molecule, the "creeper" (Fig. 4) was found in [41] to drift steadily over 1.6 hours towards the line center. The wandering of this molecule presented small discontinuous jumps in addition to the drift, suggesting a driven random walker. This drift might result from a structural relaxation of the strained crystal over a long timescale. Within the more specific model of interaction of the molecule with a domain wall [43, 62–64], the long term trend of the trajectory could be explained by a slow migration of the wall, while the random jumps would result from wall dynamics.

The dynamical degrees of freedom responsible for the observed spectral diffusion must be in thermal equilibrium with the phonon bath. An increase of temperature should therefore lead to the activation of more degrees of freedom, as well as to faster jumps. The dependence on temperature of the frequency trajectories above 3 or 4 K was easiest to follow with the fiber setup (it was not possible to follow the molecule with the lens because of the refractive index change between liquid and gaseous helium). The dependencies measured clearly show an increase both of jump rate and of the explored range in the wandering pattern, as the traces of Fig. 5 attest. The increase is particularly clear above 4 K, when librations of the molecules in the crystal start to be activated [65]. The increase of the wandering range is compatible with the activation of new degrees of freedom, which were hitherto frozen. Crude statistics for one molecule showed that the rate of jumping approximately doubled between 4.2 and 5.7 K, from 1 jump in 80 s to 1 jump in 40 s.

Finally, let us consider the question of the time scale of spectral jumps in pentacene/p-terphenyl. Frequency trajectories reveal characteristic times of the order of minutes at a few K [41]. Since this method has a limited time resolution, one may wonder if faster jumps also occur. Of course, the traces of wandering molecules show shorter times between jumps, because the number of coupled TLS's is large, but the characteristic times are still in the same range. Faster jumps, which would go unnoticed on a low time resolution, can cause additional broadening of some single molecule lines. However, such kind of information is very difficult to draw from long-term measurements. The correlation method was applied to the study of photon bunching by triplet saturation in pentacene/p-terphenyl. The correlation functions measured did not show characteristic times beyond those attributable to the triplet [66], but they were measured on stable molecules. A possible case of fast jumping is the example of the two nearby lines of [59], which very probably belonged to the same molecule, since they still appeared together after several thermal cycles up to 8 K. As the two lines appeared on the same scan, lasting a few seconds, without clear signal interruptions, the jumps had to be faster than about 0.1 s. These lines were observed to merge at a temperature of about 3 K into a single line. A possible interpretation of this merging is a motional narrowing effect: when the jump rate becomes comparable to the frequency difference between the lines, one gets a single line at the average frequency [59]. This would imply jump rates in excess of 100 MHz, much faster than those observed in the trajectories, of the order of seconds. Such fast jumping processes could occur in some defects in severely strained regions of the crystal, where the tunneling barrier for the middle ring of some molecules would be particularly depressed. Since, however, the occurence of the merging behavior was very rare (two molecules in several hundreds), the fast jumping times could not yet be confirmed by measurements with a correlator (see next section).

The study of single pentacene molecules in p-terphenyl shows the value of this method for investigating spectral diffusion in solids. The evolution of microscopic defects can be followed in real time, revealing a wide variety of possible local environments of molecules and of correlated behaviors for the molecular optical line.

This broad array of patterns is of course out of reach of more classical methods, like spectral hole-burning, which average over the many individuals in a bulk. The statistics over many molecules show a clear correlation of spectral jumps with strong static shifts of the line. This shows that molecules located near crystal defects are more prone to spectral jumps, in other words that spectral diffusion mainly originates from strained or disordered parts of the crystal, as expected. The theoretical model of Reilly and Skinner [43, 62–64] (see Section 1.5), interprets the spectral jumps by transitions of TLS's involving reorientations of the central phenyl ring of the *p*-terphenyl molecule. These TLS's would correspond to *p*-terphenyl molecules in a domain wall adopting one of the two distinct conformations belonging to either of the adjacent domains. The experimental trajectories, in particular the distribution of jump amplitudes, are compatible with a two-dimensional arrangement of independent TLS's. The jump time for this kind of defects is of the order of hundreds of seconds at a few K [62], and parameters such as the interaction between pentacene and TLS and the distance to the wall can be deduced from fits to the experimental data. Since the interaction falls off quickly with distance, only those pentacene molecules that reside within a few lattice constants of the domain walls exhibit spectral diffusion.

1.4.2.3 Other crystalline systems

Recently, Kummer, Basché and Bräuchle included terrylene in a *p*-terphenyl crystal [67]. The data on this new system are limited so far, but as with pentacene, there are 4 origins in the emission and excitation spectra. One may expect to recover the spectral jumps related to *p*-terphenyl crystal dynamics, probed by a different molecule. The much larger bulk of terrylene as compared to pentacene suggests, though, that the insertion sites differ deeply for the two molecules. Indeed, three of the terrylene sites are photo-unstable, i.e. the molecules undergo large and often irreversible jumps upon excitation (reversible jumps also occur). One of the sites (X_2) is much more photostable than the others, which allows study of the saturation of terrylene at high exciting intensity for example. The existence of photoinduced jumps, perhaps to other sites, can be interpreted by light-induced transitions of nearby TLS's in the strained neighborhood of terrylene molecules, whereas the much less strained pentacene sites are completely photostable. The walls between, *p*-terphenyl domains could also be pinned at or close to terrylene molecules, providing a large number of TLS's that could flip under optical excitation of the molecules.

Finally, one of the crystalline systems that lately appeared suitable for single molecule spectroscopy is terrylene in an *n*-hexadecane Shpol'skii matrix [68, 69], probably the first instance of a large class of systems. The excitation spectrum of terrylene in *n*-hexadecane shows at least one well defined site at 571.9 nm, together with a broader background of molecules, probably situated at crystal defects. Spectral jumps of several single molecules in this system were studied by Moerner et al. [68]. The molecules were less stable than class I pentacene molecules in *p*-terphenyl, but the jumping molecules tended to show multistate jumping between well-defined spectral positions, whereas the wandering characteristic of highly strained *p*-terphenyl crystals was not observed.

Figure 6. Frequency jumps of a terrylene molecule in the Shpol'skii matrix *n*-hexadecane. Left: Grey-scale plots of the fluorescence intensity *vs.* laser frequency in about 600 laser scans; right: deduced frequency jumps of the molecule (from Ref. 68).

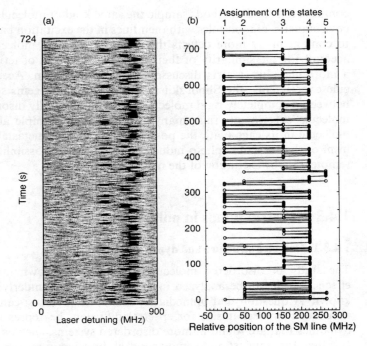

The simplest case was that of two-state jumpers. In several examples, the frequency difference between the two spectral positions lay around 150 MHz, implying the existence of a TLS at a given, rather large distance from the molecule. Varying the exciting intensity, the authors of [68] found that in most cases the jumps are light-driven: a molecule can be stable for long times at low power and starts to jump at high power. The dwell time between jumps was also measured and found to depend on the excitation intensity. These jumps are often reversible, i.e. the molecule and its associated TLS behave as a molecular switch.

There were also multistate jumpers, molecules with several preferred frequencies. For one of them, as shown in Fig. 6, at least 5 states with different frequencies could be found [68]. These 5 states could arise from interaction with at least 3 TLS's, with their flipping rates such that only 5 out of the 8 possible positions are practically detectable. Else, a real 5 state jumper could be associated to a multivalley configuration of the molecule-matrix geometry (this would correspond to a set of TLS's interacting so strongly that the TLS description loses its meaning). Using hole-burning, Attenberger and Bogner [70] found similar complex multi-well potentials, with photo-induced transitions between the deeper well and metastable ones and with slow back reactions, for the closely related system perylene in *n*-heptane.

Beside the site line at 571.9 nm, terrylene in hexadecane presents a broad background in which single molecule lines can be observed and persistent spectral holes can be burned [69]. The histogramme of single molecule linewidths compares well with the average width of the holes, extrapolated to low burning fluences, which confirms that both methods sample the same kind of molecules in similar ways. The environments of molecules with their lines in the excitation background are expected to contain many more defects than those of the site molecules of [68]. This may account for the similarity of their behavior with that of terrylene in polyethylene, a more glass-like matrix discussed in the next section. According to single molecule experiments, the Shpol'skii matrix hexadecane seems somewhat intermediate between a completely rigid molecular crystal and a fully disordered polymer. Single molecule experiments raise many questions, for example about the fitting of the molecule in its cage (with the possibility of multiple metastable states), the mechanism of the reversible photo-induced jumps and the possibility of switching a fairly remote TLS by excitation of the molecule [68].

1.4.3 Spectral jumps in polymers

1.4.3.1 Polymer structure and dynamics

The study of disordered molecular crystals has shown many spectral diffusion effects, which can be analyzed in detail because the underlying crystal structure is known. A question that immediately arises, however, is: can any one crystal, with its particular mechanisms for spectral diffusion, be representative of the kind of dynamics that are found in more disordered systems, such as glasses and polymers? It is therefore interesting to study spectral diffusion in molecular glasses because the disorder there is of a general kind, and one can expect to find the same kind of degrees of freedom that are responsible for specific low temperature properties of most other disordered solids. As discussed in the introduction, the general problem of glassy dynamics at low temperature has been tackled by several methods, involving many different physical measurements. The point of using single molecules to study glass dynamics is that they probe the local environment at nanometer level, so that all the averagings entailed by classical methods, even selective ones such as hole-burning, are avoided. A wide variety of different environments may thus be sampled, which is particularly important in strongly disordered systems. The only glassy systems used so far in single molecule spectroscopy are polymers, because, being solid at room temperature, they are easy to prepare and manipulate. The first polymer used as a matrix was polyethylene [3, 71], in which several guest molecules were known from previous hole-burning measurements to give narrow zero-phonon lines [72]. The reasons for this probably lie in the structure of this polymer, whose long saturated chains crystallize easily. The structure of polyethylene is semi-crystalline, with crystals lying in radial arrays known as spherulites [73]. The crystalline lamellae alternate with amorphous layers on a scale of 500–1000 Å. The fraction of crystallized to amorphous material, called crystallinity, is related to the density and depends on chain length and on the amount of side-chains. Since aromatic molecules such as 9-bromoanthracene and acridine are found to be either included in the amorphous regions or adsorbed at the surface of crystallites [73], the larger molecules used for spectroscopy in the red part of the spectrum are expected to be embedded in the same way. When going from molecular crystals to fully amorphous polymers and to glasses, polyethylene is a convenient intermediate step, for it retains most of the advantages of crystals (narrow lines, weak spectral diffusion and hole burning), while providing insight into the many possible degrees of freedom of glasses that are still activated at low temperatures. These include tunneling excitations known as TLS's, but also strongly anharmonic or low-frequency phonons localized in disordered regions, which might be responsible for dephasing of the optical transition. So far, single molecule studies of dynamical processes in the matrix have been limited to timescales longer than microseconds, far too long for dephasing studies. More recently, truly amorphous polymers have been used as matrices for single molecules, giving broader lines and richer dynamics than polyethylene. A brief account of current results in these systems will be given below, but detailed studies require much additional work.

1.4.3.2 First polymer system: perylene in polyethylene

The success of the fluorescence spectroscopy of pentacene in *p*-terphenyl crystal immediately suggests including pentacene in a different, disordered matrix such as polyethylene. Several attempts in this direction failed, first because of the photo-reactivity of pentacene [42] and second because the matrix can greatly influence the photophysical properties of pentacene [66]. As discussed in Section 1.2, the inter-system crossing rate from the excited singlet of pentacene strongly depends on the local site structure. Different environments around pentacene sites in *p*-terphenyl crystal lead to hundredfold increases of the triplet yield [74]. It is therefore likely that the highly flexible pentacene molecule has an unfavorable triplet yield when included in a polymer.

In 1991, Basché and Moerner included perylene in polyethylene [71]. The small and rigid perylene molecule has good emission and triplet properties, but absorbs in a difficult spectral region, around 445 nm. Because the spectral jumps we discuss here are mainly consequences of matrix dynamics, we must shortly discuss sample preparation. The samples of [71] were made from low-density polyethylene (crystallinity 25%), doped at low concentration with perylene, and were quickly quenched from the melt to liquid nitrogen temperature to reduce light scattering. The thin films thus obtained were 10 to 20 μm thick. The polymer structure is thus expected to be dominantly amorphous.

The study of the excitation lines of several single perylene molecules showed distinctive features, characteristic of a polymer matrix. Instead of a well defined homogeneous width close to the natural width, as in pentacene in *p*-terphenyl, the single molecule widths are broadly distributed between 50 and 150 MHz. All widths are considerably larger than the lifetime-limited width of perylene, 23 MHz. Since the laser intensity was too weak to saturate the perylene transition, the broadening was attributed to dephasing and/or spectral diffusion by polymer dynamics. The different widths reflect the different distributions of defects around each single molecule. No correlation was found between the width of the lines and their spectral position in the inhomogeneous band. In addition to the broadening, which is an indirect proof for small spectral jumps on a timescale shorter than that of the measurement (seconds to minutes), large spectral jumps were observed, as in the pentacene in *p*-terphenyl system. An important difference from the latter system is that here, at least part of the spectral jumps are light-driven, i.e. they are examples of non-photochemical spectral hole-burning processes. The photo-induced origin of an irreversible spectral jump is impossible to ascertain in any given case, unless lengthy statistical analyses are carried out. However, spectral jumps in polyethylene are often reversible for low exciting power. Reversible photo-induced jumps lead to spontaneous hole filling in bulk samples [39]. The back reaction provides a simple method to prove the light-driven nature of the jumps: since the line spontaneously jumps back to its original position after a while, it is possible to establish that the average jump rate increases with exciting laser power. It is also possible to keep track of the burn times (i.e. the time a molecule withstands excitation before jumping), and to plot their histogram. The distribution of burning times found is compatible with a single exponential [42], as expected if burning events are independent and random. Such a histogramme of burning times is closely related to the kind of fluorescence intensity correlation functions that will be discussed below. The results of [42] showed for the first time the wide variety of local environments; furthermore, single reversible hole-burning events were characterized and studied on single molecules. The authors of [42] also raise the question of a possible correlation between the size and the timescale of the jumps. This problem is difficult to address: because the many small jumps from the remote background contributing to spectral diffusion cannot be resolved individually, their time constants are unknown (contrary to the few large jumps arising from close molecular surroundings).

1.4.3.3 Excitation lineshapes and widths

Terrylene in polyethylene

In 1992, another polymeric system was proposed for single molecule studies in a more convenient part of the spectrum, terrylene in polyethylene [3]. Terrylene is a heavier analog of perylene, containing 3 naphthalene rings, which absorbs in the yellow part of the spectrum (absorption maximum at 570 nm in polyethylene) and possesses a good emission yield [75]. The polyethylene used in [3] was different from that of [42]. It had high density, a crystallinity of 75%, and was not quenched, which means that the density of defects was probably less than in the low density quenched polymer. We have seen that most aromatic molecules should lie in the amorphous parts or on crystallite surfaces, but the difference in crystallinity must be kept in mind when comparing the observations from Ref. 3 and 42. Just as with perylene, single terrylene molecules present different shapes and widths as shown in Fig. 7 [44, 76]. The Lorentzian shape (within experimental noise) was found most often, but asymmetrical profiles, shoulders or split lines were common. The fraction of split lines measured as doublets amounted to a few percent, of which a significant part probably arose from chance coincidences of the lines of two different molecules. In a few cases, quadruplets (4 lines) were seen and proved to belong to the same molecule when they disappeared together after a large spectral jump [77], whereas triplets were invariably chance coincidences between a doublet and a single line. Occurence of doublets and quadruplets are strong indications of coupling of the molecule to

Figure 7. Sample of single-molecule resonances of differing widths for the system terrylene in polyethylene at 1.8 K (from Ref. 76).

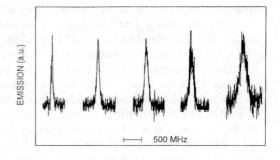

Figure 8. Distribution of the widths at half maximum of around 200 molecules of terrylene in polyethylene. The lower cut-off corresponds to the lifetime limited width, indicating molecules by chance isolated from tunneling systems, the peak to molecules in an average environment and the tail at large widths to molecules probably influenced by more than the average number of tunneling systems.

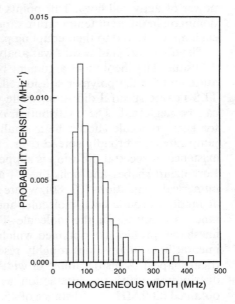

TLS's, as are the two-state trajectories of [42]. The dynamics within doublet molecules will be discussed in the next section.

The distribution of single molecule excitation linewidths was recorded and discussed in [76]. The histogramme of widths for a few hundreds molecules is presented in Fig. 8, with widths ranging from 50 to a few hundreds of MHz. The widths were measured by scanning the resonance at low power to avoid hole-burning, with intensities much less than the saturation intensities of terrylene [78, 79], so that power broadening by saturation of the two- or three-level molecular system was negligible (However, laser irradiation could have induced jumps during the measurement, leading to additional spectral diffusion which will be neglected in the present discussion). The histogramme of widths presents a sharp onset on the low side, and a slow decrease toward large widths. The cutoff at small widths is compatible with the expected natural width of terrylene (about 40 MHz, as deduced from lifetime measurements [68]), implying that some molecules are nearly unperturbed, i.e. they have no defects in their close micro-environments. Most molecules have widths of less than 200 MHz, with a most represented width of 80 MHz. The absence of widths larger than 400 MHz can be ascribed to the fact that close TLS's will lead to line splitting rather than to broadening. The strongly interacting TLS's leading to split lines are therefore eliminated from the width statistics. The widths of a few molecules were also studied in [76] as functions of temperature. The difficulty of such studies is that the probability of large irreversible jumps quickly increases at higher temperatures. Just as the widths are distributed according to the molecular environment, their temperature dependences also strongly differ from molecule to molecule. In the domain between 1.7 K and 4 K, the width can vary linearly or faster, with power or activated laws. This points to different broadening mechanisms, that can dominate in different temperature ranges, according to the particular defects around each molecule and to their coupling parameters.

The distribution of widths was analyzed in [76, 77] with a simple model of spectral diffusion. The theoretical argument behind this assumption is that, at the temperature and for the polymer considered, the main source of dynamics is TLS's. Slow TLS's cause spectral diffusion, while very fast TLS's are motionally narrowed and can be neglected. The contribution of TLS's with characteristic times around T_2 for each molecule should be a small fraction of the total number (because TLS jump rates are broadly spread over many decades) and can also be neglected. The existence of spectral diffusion is supported by the experimental observation of non-Lorentzian shapes and split lines. Further support for this model is provided by numerical simulations [77, 80], where particular configurations of TLS's are picked at random around each molecule, and an inverse cubic power dependence on distance is assumed for the molecule–TLS interaction. By simulating the long-term lineshape, profiles are obtained which qualitatively agree with the shape of experimental lines (many lines roughly resemble Lorentzians, with a few odd, asymmetrical shapes). A histogramme of widths can also be built and fitted to the observed distribution using the interaction as the only adjustable parameter. The value obtained (200 MHz at a distance of 45 Å) is consistent in order of magnitude with the deformation potential of TLS's and the pressure shift of single molecule lines (about 1 GHz/bar). Most single molecules are therefore mainly broadened by remote TLS's, which but very little affect the microscopic structure of the matrix around the molecule. This agrees with the observation that spectral jumps of a few GHz hardly change the molecular dipole moment difference between ground and excited states [3].

Lineshapes and widths in other polymers

The first truly amorphous polymer in which single molecules were detected was polyisobutylene [81]. The guest molecule was tetra-*t*-butylterrylene, which was

Figure 9. Distribution of linewidths of molecules of terrylene in polystyrene at 1.8 K. Sample lineshapes are shown above the histogram. The peak of the histogramme agrees well with the "homogeneous" linewidth deduced from hole-burning on the same system (from Ref. 69).

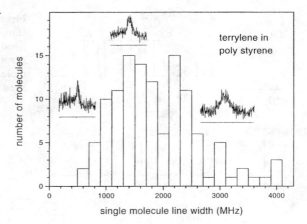

also studied in polyethylene. The observations for both systems were very similar to those of terrylene in polyethylene (this again suggests that terrylene molecules are chiefly included in the amorphous regions of polyethylene), except for photo-induced spectral jumps, which were more frequent in polyisobutylene. For example, the distributions of widths shows similar cutoffs and wings to those in [76], with somewhat larger widths in polyisobutylene.

Terrylene was also included in three completely amorphous polymers: polyvinylbutyral, polymethylmethacrylate and polystyrene [69]. Again, single molecule lines were recorded and found to vary greatly in shape and width. In all three matrices, lines were much broader than in polyethylene (the average width was 0.6 GHz in polyvinylbutyral, 1.0 GHz in polymethylmethacrylate and 1.8 GHz in polystyrene). The histogrammes of widths were much more symmetrical than in polyethylene (see the example that of polystyrene in Fig. 9), which does not agree with the simple model explained above, based on spectral diffusion induced by randomly placed TLS's. Assuming that the density of TLS's is roughly the same in all amorphous molecular solids [7, 11], the only parameters left to adjust the theory to the experimental histogrammes are the interaction and the minimal approach distance between molecule and TLS's. The latter radius has to be unreasonably large to obtain nearly symmetrical or Gaussian distributions. This means that the model is probably not correct for those polymers, where the average width is much larger than the natural width. Possible explanations for this failure include among others the existence of excitations other than TLS's (e.g. quasi-local phonons), and a strong interaction between TLS's.

Photo-induced processes in these disordered matrices were much more frequent than in polyethylene, making the observation of single molecules more difficult, but conventional hole-burning easier. The homogeneous width deduced from hole-burning by extrapolating to low burning fluences compared very well to the single molecule average width in all three polymers, confirming the equivalence of the two types of measurements. The saturation of the burning process, which very soon leads to broadening of spectral holes, obviously does not affect single molecule lines.

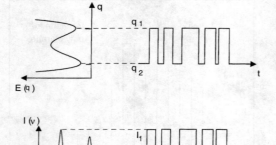

Figure 10. Illustration of how changes of state of a tunneling system in the neighbourhood of a molecule cause a "random telegraph" fluorescence signal by shifting the molecule in and out of resonance with the fixed laser frequency.

1.4.3.4 Dynamical properties

The ideal way to characterize spectral diffusion and spectral jumps of a single molecule would be to follow its optical line in the spectrum with all time constants ranging from T_2 (nanoseconds or shorter). Unfortunately, finding the resonance frequency of a molecule with reasonable accuracy demands at least several tens or hundreds of photons, which limits the accessible timescales of trajectories to milliseconds or longer (most trajectories reported so far [41] had a time resolution of seconds rather than milliseconds). The intensity correlation method provides dynamical information on much shorter timescales. When the emitting molecule drifts or jumps through the spectrum, the intensity of fluorescence excited by a laser at a fixed frequency fluctuates (Fig. 10). A convenient way of keeping track of these fluctuations over some long period of time is to record the auto-correlation function of the fluorescence intensity [82] (cf. Section 1.1). The correlation function records the number of photon pairs separated by τ over some long integration time T_{exp} as a function of τ. The advantage of this kind of measurement is that even very slow processes (provided they are faster than T_{exp}) will contribute to the correlation function. The full correlation function will carry information about processes with characteristic times ranging from the photomultiplier's dead time (a few nanoseconds) to minutes or longer, according to the duration of the acquisition. A convenient way of recording and displaying such a broad range of times is to use a logarithmic correlator. This device is particularly useful for the study of systems in which timescales are broadly spread out, like glasses and polymers at low temperatures.

The correlation method was applied to a thorough study of spectral diffusion of single terrylene molecules in polyethylene [44, 76], and later on to other systems [81]. Once a single molecule line was identified in the excitation spectrum, the laser was brought into resonance with the line, and the correlation function of the emitted fluorescence was recorded. For most single molecules studied (around 80%), no clear correlation appeared between 1 μs and 100 s, i.e. the contrast of the correlation was weaker than experimental noise. This is in general agreement with bulk studies of

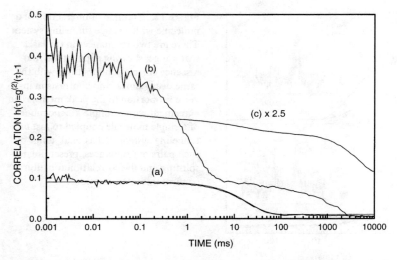

Figure 11. The auto-correlation function of the fluorescence intensity reveals the characteristic times of the "random telegraph" signal in Fig. 10. Here, three examples of correlation functions of different terrylene molecules in polyethylene at 1.8 K illustrate (note the logarithmic time axis): (a) an exponential decay (with fit, smooth line); (b) a bi-exponential decay and (c) a decay with many timescales. Such different behaviours may reflect molecules coupled strongly to one, two or many tunneling systems.

other chromophores by hole burning [83], indicating the absence of a significant spectral diffusion in polyethylene. Let us remark here that the absence of hole broadening by spectral diffusion does not necessarily imply the absence of any TLS: If each molecule is coupled to one or to very few TLS's, a spectral hole will not broaden with time, as long as a significant fraction of the line remains at the original frequency. Its area will simply decrease. The correlation method, however, will show the intensity fluctuations due to a single TLS. By selecting single molecules, it is thus possible to focus on the remaining 20% of molecules, which do show structure in their correlation function. The few examples shown in Fig. 11 show that the correlation functions measured differed greatly in shapes, contrasts and characteristic times, an illustration of the variety of molecular environments and timescales in a polymer at low temperature. Among the many correlation patterns observed, a fairly common one was the single exponential step, which was attributed to coupling of the molecule to a single TLS, on the basis of theoretical arguments developed hereafter.

The simplest case of spectral diffusion is that of a single molecule coupled to a single TLS in its neighborhood. Upon interaction with acoustical phonons, the TLS will jump from one configuration to the other, thereby changing the molecular optical transition frequency. The fluorescence intensity will therefore fluctuate according to the jumps, giving rise to two lines in the excitation spectrum and to a correlation function:

$$g^{(2)}(\tau) = 1 + k_1 k_2 [(I_1 - I_2)/(k_2 I_2 + k_1 I_1)]^2 \exp[-(k_1 + k_2)\tau].$$

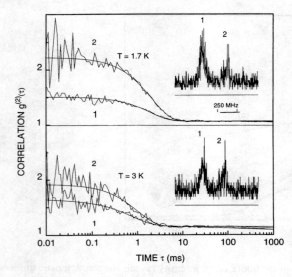

Figure 12. Example of interaction of a molecule with a single tunneling system. There are two resonances in the laser scan range, showing exponential fluorescence correlation functions with the same decay times, and contrasts in inverse proportion to the peak intensies, in agreement with a simple kinetic model of a single molecule coupled to just one tunneling system. Occasional jumps of such pairs of resonances, preserving the splitting and the correlation function confirm this interpretation.

$k_1(k_2)$ are the inverse lifetimes and $I_1(I_2)$ the fluorescence intensities in well 1(2). This simple model predicts the occurence of split lines presenting the same correlation time-constant and contrasts inversely proportional to the intensity of the lines, as observed experimentally (see Fig. 12; in the case studied in [76], the attribution of the two lines to the same single molecule was further confirmed by occasional jumps of both lines together between two frequency positions, a few GHz apart). Such observations with two well defined and separated lines strongly support the TLS model.

The situation corresponding to the simple model discussed above occurs rarely in a real disordered system. To begin with, molecules are always coupled to far-away TLS's, which will broaden their lines, or cause drifts on a broad range of timescales. Moreover, more than one TLS can be close enough to the molecule to cause, for example, two splittings and a quadruplet of lines, with particular intensity ratios [77]. However, such cases occur fairly seldom. If a TLS is too close to the molecule, the components are too far apart to be recorded in the same laser scan, or, with higher probability, if a TLS is too remote, the line is not split completely. This does not prevent the characteristic time of the TLS from appearing – with a low contrast – in the correlation function, if the frequency jump is not much smaller than the linewidth.

Numerical simulations were performed in the authors' group in order to better understand experimental observations [77, 80]. They are based on minimal assumptions about the system: a random distribution of TLS's in space around the molecule, an interaction decreasing like the inverse cubic distance between molecule and TLS, and a broad spread of asymmetries and jumping times due to tunneling dynamics and disorder. No correlation was assumed between jumping rates, asymmetries and distances from the molecule. Simulated lineshapes and correlation functions are qualitatively similar to experimental data (Fig. 13). They confirm that the isolation of a single TLS in the correlation function is possible, even when several

Figure 13. Stochastic simulation of the fluorescence intensity (top) and frequency (middle) correlation functions for a model of spectral diffusion in a glass. The molecule is coupled to a three dimensional distribution of tunneling systems with distributed microscopic parameters. The bottom panel shows the timescale (horizontal) and amplitude of the frequency jumps (relative to the lifetime limited linewidth, vertical scale). The inset in the top panel shows the lineshape and the lifetime limited linewidth as a small bar (from Ref. 77).

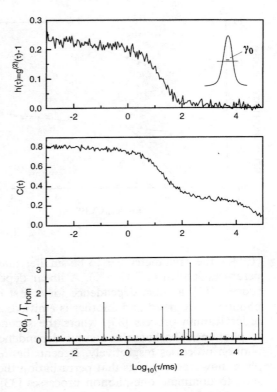

other systems are coupled to the molecule and many frequencies are explored as time passes. It is therefore impossible to conclude from a multiplicity of explored frequencies and from a single correlation time that all coupled TLS's have the same time constant [81]. Simulated frequency trajectories present many small jumps and fewer large jumps corresponding to nearby defects, resulting in a picture similar to the Levi flights or walks of [84].

Once a single exponential contribution has been isolated in the correlation function of a single molecule, it is thus safe to assume that it arises from a single TLS in the neighborhood of that molecule. The characteristic time of the correlation gives the sum of jumping rates between the wells, which can be further studied as a function of external parameters such as temperature or light intensity. We first discuss temperature effects. For all TLS's studied so far, the jumping rates were found to increase with temperature, but the dependences were relatively weak, as expected if phonon-assisted tunneling is responsible for the jumps (see a few examples displayed in Fig. 14). The most striking result of the study of several TLS's is the presence of different temperature laws, either algebraic or of the Arrhenius type [44]. The algebraic laws were either linear or cubic. These dependences are expected from a simple perturbation theory of the interaction of a TLS with acoustic phonons, if the asymmetry is much smaller than the thermal energy. Indeed, because TLS's were selected experimentally by first studying correlation functions at the lowest temperature

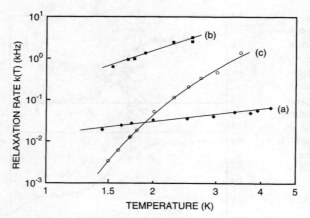

Figure 14. Temperature dependence of the decay rates of the fluorescence correlation (tunneling rate of the tunneling system coupled to the probe molecule) of three different molecules of terrylene in polyethylene. The smooth lines are fits showing proportionality to T (a), to T^3 (b) and to an activation law (c), predicted to be possible for different modes of coupling of the tunneling system to phonons.

(1.7 K), the asymmetry has to be smaller than thermal energy in the range of temperatures studied (1.7 to 5 K). A linear dependence is obtained for a one-phonon process [11], a cubic dependence for a first order Raman process [85], where one phonon is absorbed and another is emitted, and a seventh power law for a second order Raman process [85], where the second derivative of the potential couples to phonons. The linear and cubic dependences were interpreted as one- and two-phonon processes respectively. Recent theoretical work on the TLS–phonon coupling, however, suggests that perturbation theory fails as soon as Raman processes start to dominate one-phonon processes [33]. Should this hold true, the interpretation of the cubic dependences would have to be revised. In one case [44], an Arrhenius dependence was obtained for the jump rate over a small range of temperatures (between 1.7 and 5 K), but which corresponded to a thousandfold increase in rate. A purely classical activation probably may be excluded, since the prefactor would have to be too small. Still, several mechanisms involving tunneling could explain such an activation. A first possibility involves tunneling via the excited states of either well, which first needs activation to reach these states. Another mechanism could be a two-phonon Raman process involving a quasi-local mode. Another still could involve correlated tunneling along one coordinate and classical activation along another one, corresponding to a coupling to slow motion. This last interpretation was discussed by Kagan in a review of coherent and incoherent tunneling processes [86].

The exciting laser intensity is another parameter which can be easily varied to check its influence on the jumping rate of TLS's. By using the contrast of the correlation function, and assuming that only one TLS is strongly coupled to the molecule, it is possible to deduce both rates between wells, whereas the correlation decay rate only gives the sum of the rates. Photo-induced jumps have been observed in all polymers studied so far [42, 69, 81], although an unspecified fraction of molecules seems almost insensitive to light. Examples have been given in [76] of a strong increase of the contrast and rate with light intensity and of a nearly insensitive molecule. In the cases studied, the light-induced effect is compatible with the simplest model, where the rate of transition to the metastable burned state is proportional to the excited state population [76, 77]. Usually, jump rates [42] as well as the range of spectral diffusion [78] (or the linewidth measured over long integration times) increase more or less with intensity. Rather often, the shape of the correlation function also changes at high intensity, for example from a single exponential decay, to multi-exponential, or to a roughly linear decrease on a logarithmic timescale [81]. These observations can be interpreted as the onset of light-induced jumping of more and more TLS's as the power is raised. This phenomenon is well known in bulk systems where non photochemical hole burning is often accompanied by a back reaction (spontaneous hole filling [39]). Several mechanisms have been proposed to explain photo-induced jumps; for example, tunneling may be easier in the excited state than in the ground state because of a lower or narrower barrier [87]. In the Born–Oppenheimer picture, a non-radiative transition from the excited state leads to a highly excited vibrational level of the ground state, which may allow exploration of the ground state potential hypersurface during its short lifetime, and eventually relaxation to a new well. A closely related mechanism assumes that heat released by a non-radiative transition and mediated by matrix phonons may activate nearby TLS's of the matrix. The observation of photoinduced jumps of small amplitude in polymers and hexadecane [68] (from a few hundreds of MHz to a few GHz) shows that even remote tunneling systems can switch upon optical excitation of the chromophore, which suggests a new, less local view of non photochemical hole-burning mechanisms.

Tunneling systems should in principle be sensitive to other external perturbations like an electric field or hydrostatic pressure. Attempts to observe changes in the correlation function by applying modest electric fields have been unsuccessful so far. Taking into account the weakness of dipole moment changes of TLS's in a non polar polymer, this is not very surprising. Most TLS's are also insensitive to jumps of nearby TLS in the surroundings of the same molecule. Occasionally, however, the correlation function [76] or the frequency trajectory [68] can change deeply after a slow spectral jump, which shows that neighbor TLS's can sometimes strongly interact.

1.4.4 Conclusions

1.4.4.1 Main results

Single molecule experiments for the first time gave evidence for the discrete nature of spectral jumps responsible for spectral diffusion. In amorphous matrices, and also in disordered crystals, these experiments demonstrated that a crucial contribution to spectral diffusion originates from the close neighborhood of the molecule, which leads to rejection of models based on continuous drifts of the transition frequency, where remote regions of the sample would play the leading part. The wide variety of local environments in a glass or in a disordered crystal is responsible for the distribution of molecular parameters (there is no such thing as 'the' homogeneous optical width in a glass, since each molecule has a different linewidth) and of several qualitatively different behaviors. Most of the results obtained so far about lineshapes, trajectories or correlation functions are consistent with the simplest model of a low-temperature glass, the standard TLS model. However, the occasional observation of more complex behaviors suggests that refinements will be needed, for instance in order to explain the presence of a long-term trend in some trajectories [41], the interaction of TLS's [76], the occurence of multistate jumping that cannot be described by a set of TLS's [68], or gradual intensity changes that take place without frequency drift [78].

Jumps between two spectral positions were found to vary considerably both in amplitude and timescale. The jump size ranges from smaller than the homogeneous linewidth (tens of MHz or less) to several tens of GHz and probably more, while the timescales are on the order of minutes for pentacene in *p*-terphenyl, or range from shorter than microseconds to minutes and beyond in polymers. This is consistent with a random distribution of flipping two-state defects around the molecule, with a wide spread of jumping rates, as expected for tunneling systems. All these features are related to the well-known spectral diffusion of optical lines in bulk systems observed by spectral hole-burning and photon echoes. The main conclusions that can be drawn concerning spontaneous jumps are:

i) The main features of the spectral jumping behaviors investigated so far can be described by flipping of unspecified effective particles between two nearly degenerate positions (asymmetrical double well). The order of magnitude of the flipping rates and their temperature dependence unambiguously warrant an assignement to tunneling assisted by acoustic phonons.

ii) Spectral diffusion trajectories in *p*-terphenyl crystals are well understood if the molecule is coupled to a two-dimensional arrangement of TLS's. This structure may be interpreted as a domain wall between domains of the crystal with different symmetries [43]. Single molecule spectral trajectories will give access to a wealth of information about the tunneling systems when more data become available. Since the crystal structure is well-known, detailed comparison with theory will be possible.

iii) In the more difficult problem of tunneling in amorphous systems, a good agreement was obtained between experimental data and the standard TLS model, as used for example in numerical simulations. Spectral diffusion can be seen as a random walk in the set of frequencies specified by the states of all coupled TLS's. Individual TLS's can be isolated and studied independently over minutes or hours, confirming in several cases model assumptions, but more complex behavior also occurs. A good agreement with simulations was obtained in polyethylene, a matrix where spectral diffusion is weak or absent. In the fully amorphous polymers where lines are broader, the agreement with the TLS model seems to be less good. However, more spectral and dynamical data are needed before specific hypotheses of the model can be questioned.

Single-molecule lines also undergo photoinduced jumps, consecutive to absorption of a photon by the probe molecule. Several effects including persistent spectral hole-burning, light induced spectral diffusion, hole filling, and laser induced hole filling are well known in bulk solutions. All the corresponding processes have been observed with single molecules, but more detailed information will become available from the study of discrete jumps in each case (which is usually impossible in bulk systems). For example, a very important question in non photochemical hole-burning is that of the nature of the (extrinsic) TLS's involved. It was thought that extrinsic systems were situated close to the molecule, and therefore different from those responsible for spectral diffusion. Single molecule studies, however, show that spontaneous jumps (that occur without light applied) can be driven by excitation of the molecule [76], and that even fairly remote TLS's (giving rise to small jumps) can be light-driven [68]. Careful study of such cases may help decide between different mechanisms of photoinduced jumping (e.g. tunneling in the excited state, or activation by heat released in non-radiative transitions).

In the study of microscopic quantities and processes such as those responsible for spectral diffusion, single molecules give access to the full distribution of parameters, instead of the averages provided by usual bulk methods. Since data obtained from single molecules are free from any averaging, they will tend to show more detail (which can get blurred or lost in bulk spectra) and they can be compared to extremely simple theoretical models, where no average need be taken. This is a big advantage for the investigation of complex systems. Single-molecule measurements are difficult and lengthy, so improvement of statistics by faster data acquisition and processing will be a very important next step in this field. If several quantities are measured on the same molecule, higher-order statistics can also be obtained. For example, correlations in the distributions of two quantities can give hints as to the mechanisms by which both quantities are affected.

1.4.4.2 New directions

In the standard TLS model, their spatial distribution is random. In real samples, TLS's might be more concentrated around defects or particularly disordered regions of the sample. Microscopy of single molecules, especially with near-field resolution might help correlate TLS's with defects or structures in the sample. In samples with artificial structures, the influence of external parameters on spectral jumps and TLS's could be studied.

The tunneling particle in glasses is thought to be a set of atoms undergoing correlated rearrangement from one configuration to another. In the case of *p*-terphenyl crystals, a central phenyl ring of the terphenyl molecule flips from one position to the other with respect to the extreme rings. It seems very promising to study tunneling of lighter particles such as protons or electrons, over larger distances (in order to get tunneling rates in the experimental window). Artificial structures possessing multiple wells, such as synthesized molecules or nanofabricated structures could be studied in this way by distributing single probe molecules in their neighborhood.

The intensity auto-correlation method has long been in use at high temperature to study dynamics in solutions [88]. Intensity fluctuations can arise from spectral jumps or changes, from rotational or translational diffusion with respect to the exciting beam, and from any process which can modulate the emitted intensity. Since correlation of single molecule fluorescence works at liquid helium temperature as well as at room temperature (see Section 2.1), it probably can cover the whole intermediate temperature range, where many different processes occur with a broad range of activation energies. Single molecules provide unique opportunities to study these processes at a truly nanoscopic scale.

References

[1] K. K. Rebane, *Impurity Spectra of Solids*, Plenum Press, New York, 1970.
[2] U. P. Wild, F. Güttler, M. Pirotta, and A. Renn, *Chem. Phys. Lett.* **193**, 451, (1992).
[3] M. Orrit, J. Bernard, A. Zumbusch, and R. I. Personov, *Chem. Phys. Lett.* **196**, 595, (1992).
[4] M. Croci, H.-J. Müschenborn, F. Güttler, A. Renn, and U. P. Wild, *Chem. Phys. Lett.* **212**, 71, (1993).
[5] J. Joffrin and A. Levelut, *J. Phys, Paris* **36**, 811, (1975).
[6] J. L. Black and B. I. Halperin, *Phys. Rev.* **B16**, 2879, (1977).
[7] J. F. Berret and M. Meissner, *Z. Phys. B* **70**, 65, (1988).
[8] A. Oppenländer, Ch. Rambaud, H. P. Trommsdorff, and J. C. Vial, *Phys. Rev. Lett.* **63**, 1432, (1989).
[9] W. Press, *Single Particle Rotations in Molecular Crystals*, Springer Tracts in Modern Physics, **92**, Springer, Berlin, 1981.
[10] C. Hartmann, M. Joyeux, H. P. Trommsdorff, J. C. Vial, and C. von Borczyskowski, *J. Chem. Phys.* **96**, 6335, (1992).
[11] W. A. Phillips (ed.), *Amorphous Solids: Low-Temperature Properties*, Springer, Berlin, 1981.
[12] F. H. Stillinger and T. A. Weber, *Science* **225**, 983, (1984).
[13] W. A. Phillips, *J. Low Temp. Phys.* **7**, 351, (1972).
[14] P. W. Anderson, B. I. Halperin, and C. M. Varma, *Philos. Mag.* **25**, 1, (1972).
[15] H. de Vries and D. A. Wiersma, *Phys. Rev. Lett.* **36**, 91, (1976).
[16] H. W. H. Lee, F. G. Patterson, R. W. Olson, D. A. Wiersma, and M. D. Fayer, *Chem. Phys. Lett.* **90**, 172, (1982).
[17] B. M. Kharlamov, R. I. Personov, and L. A. Bykovskaya, *Opt. Commun.* **12**, 191, (1974).
[18] A. A. Gorokhovskii, R. K. Kaarli, and L. A. Rebane, *Zh. Eksp. Teor. Fiz. Pis'ma* **20**, 474, (1974) (in Russian).
[19] R. I. Personov, in *Spectroscopy and Excitation Dynamics of Condensed Molecular Systems*, (Eds. V. M. Agranovich and R. M. Hochstrasser, North Holland, Amsterdam, 1983, Chapter 10).
[20] R. M. Macfarlane and R. M. Shelby, *J. Lumin.* **36**, 179, (1987).
[21] W. E. Moerner (ed.), *Persistent Spectral Hole-Burning: Science and Applications*, Springer, Berlin, 1988.
[22] J. M. Hayes, R. P. Stout, and G. J. Small, *J. Chem. Phys.* **74**, 4266, (1981).
[23] H. P. H. Thijssen, A. I. M. Dicker, and S. Völker, *Chem. Phys. Lett.* **92**, 47, (1982).
[24] W. H. Hesselink and D. A. Wiersma, *J. Chem. Phys.* **73**, 648, (1980).
[25] J. Friedrich, H. Wolfrum, and D. Haarer, *J. Chem. Phys.* **77**, 2309, (1982).
[26] M. Berg, C. A. Walsh, L. R. Narasimhan, K. A. Littau, and M. D. Fayer, *J. Chem. Phys.* **88**, 1564, (1988).
[27] H. C. Meijers and D. A. Wiersma, *Phys. Rev. Lett.* **68**, 381, (1992). *J. Chem. Phys.* **101**, 6927, (1994).
[28] R. G. Palmer, D. L. Stein, E. Abrahams, and P. W. Anderson, *Phys. Rev. Lett.* **53**, 958, (1984).
[29] A. Heuer and R. J. Silbey, *Phys. Rev. Lett.* **70**, 3911, (1993). *Phys. Rev.* **B49**, 1441, (1994).
[30] S. Hunklinger and M. V. Schickfus, in *Amorphous Solids: Low Temperature Properties*, (Ed. W. A. Phillips), Springer, Berlin, 1981.
[31] B. Golding and J. E. Graebner, in *Amorphous Solids: Low Temperature Properties*, (Ed. W. A. Phillips), Springer, Berlin, 1981.
[32] K. Kassner and R. Silbey, *J. Phys.* **B1**, 4599 (1989).
[33] P. Neu and A. Würger, *Europhys. Lett.* **27**, 457, (1994). *Europhys. Lett.* **29**, 561, (1995).

[34] P. Reineker and K. Kassner, in *Optical Spectroscopy of Glasses*, (Ed. I. Zschokke), Reidel, Dordrecht, 1986, p. 65.

[35] P. W. Anderson, *J. Phys. Soc. Jap.* **9**, 316, (1954).

[36] R. Kubo, in *Fluctuations, Relaxation and Resonance in Magnetic Systems*, (Eds. D. Ter Haar, Oliver, and Boyd), Edinburgh, 1962.

[37] L. Shu and G. J. Small, *J. Opt. Soc. Am.* **B9**, 724, (1992).

[38] J. M. Hayes, R. Jankowiak, and G. J. Small, in *Persistent Spectral Hole-Burning: Science and Applications*, (Ed. W. E. Moerner), Springer, Berlin, 1987, p. 153.

[39] B. L. Fearey and G. J. Small, *Chem. Phys.* **101**, 269, (1986).

[40] M. Orrit and J. Bernard, *Phys. Rev. Let.* **65**, 2716, (1990).

[41] W. P. Ambrose, T. Basché, and W. E. Moerner, *J. Chem. Phys.* **95**, 7150, (1991).

[42] Th. Basché, W. P. Ambrose, and W. E. Moerner, *J. Opt. Soc. Am.* **B9**, 829, (1992).

[43] P. D. Reilly and J. L. Skinner, *J. Chem. Phys.* **102**, 1540, (1995).

[44] A. Zumbusch, L. Fleury, R. Brown, J. Bernard, and M. Orrit, *Phys. Rev. Lett.* **70**, 3584, (1993).

[45] J. L. Baudour, Y. Delugeard, and H. Cailleau, *Acta Cryst.* **B32**, 150, (1976).

[46] J. L. Baudour, *Acta Cryst.* **B47**, 935, (1991).

[47] J. L. Baudour, H. Cailleau, and W. B. Yelon, *Acta Cryst.* **B33**, 1773, (1977).

[48] C. Ecolivet, B. Toudic, and M. Sanquer, *J. Chem. Phys.* **81**, 599, (1984).

[49] H. de Vries and D. A. Wiersma, *J. Chem. Phys.* **70**, 5807, (1979).

[50] C. Kryschi, H. C. Fleischhauer, and B. Wagner, *Chem. Phys.* **162**, 485, (1992).

[51] H. C. Fleischhauer, C. Kryschi, B. Wagner, and H. Kupka, *J. Chem. Phys.* **97**, 1742, (1992).

[52] R. W. Olson and M. D. Fayer, *J. Phys. Chem.* **84**, 2001, (1980).

[53] T. Basché, S. Kummer, and C. Bräuchle, *Chem. Phys. Lett.* **225**, 116, (1994).

[54] J. Köhler, A. C. J. Brouwer, E. J. J. Groenen, and J. Schmidt, *Chem. Phys. Lett.* **228**, 47, (1994).

[55] L. Fleury, Ph. Tamarat, B. Lounis, J. Bernard, and M. Orrit, *Chem. Phys. Lett.* **236**, 87, (1995).

[56] M. Orrit and J. Bernard, *Mod. Phys. Lett.* **B5**, 747, (1991).

[57] W. E. Moerner and W. P. Ambrose, *Phys. Rev. Lett.* **66**, 1376, (1991).

[58] J. B. W. Morsink, T. J. Aartsma, and D. A. Wiersma, *Chem. Phys. Lett.* **49**, 34, (1977).

[59] H. Talon, L. Fleury, J. Bernard, and M. Orrit, *J. Opt. Soc. Am.* **B9**, 825, (1992).

[60] W. P. Ambrose and W. E. Moerner, *Nature* **349**, 225, (1991).

[61] W. M. Itano, J. C. Bergquist, and D. J. Wineland, *Science* **237**, 612, (1987).

[62] P. D. Reilly and J. L. Skinner, *Phys. Rev. Lett.* **71**, 4257, (1993).

[63] P. D. Reilly and J. L. Skinner, *J. Chem. Phys.* **101**, 959, (1994).

[64] P. D. Reilly and J. L. Skinner, *J. Chem. Phys.* **101**, 965, (1994).

[65] T. E. Orlowski and A. H. Zewail, *J. Chem. Phys.* **70**, 1390, (1979).

[66] J. Bernard, L. Fleury, H. Talon, and M. Orrit, *J. Chem. Phys.* **98**, 850, (1993).

[67] S. Kummer, T. Basché, and C. Bräuchle, *Chem. Phys. Lett.* **229**, 309, (1994).

[68] W. E. Moerner, T. Plakhotnik, T. Irngartinger, M. Croci, V. Palm, and U. P. Wild, *J. Phys. Chem.* **98**, 7382, (1994).

[69] B. Kozankiewicz, J. Bernard, and M. Orrit, *J. Chem. Phys.* **101**, 9377, (1994).

[70] T. Attenberger and U. Bogner, in *Technical Digest on Persistent Spectral Hole-Burning: Science and Applications*, Optical Society of America, Washington D.C., 1991, **vol 16**, p. 248.

[71] Th. Basché and W. E. Moerner, *Nature*, **355**, 335, (1992).

[72] H. P. H. Thijssen and S. Völker, *J. Chem. Phys.* **85**, 785, (1986).

[73] P. J. Phillips, *Chem. Rev.* **90**, 425, (1990).

[74] H. de Vries and D. A. Wiersma, *J. Chem Phys.* **70**, 5807, (1979).

[75] A. Bohnen, K. H. Koch, W. Lüttke, and K. Müllen, *Angew. Chem. Int. Ed. Engl.* **29**, 525, (1990).

[76] L. Fleury, A. Zumbusch, M. Orrit, R. Brown, and J. Bernard, *J. Luminescence* **56**, 15, (1993).

[77] L. Fleury, Thèse, Université Bordeaux I, 1995.

[78] P. Tchénio, A. B. Myers, and W. E. Moerner, *J. Lumin.* **56**, 1, (1993).

[79] T. Plakhotnik, W. E. Moerner, V. Palm, and U. P. Wild, *Opt. Comm.* **114**, 83, (1995).

[80] L. Fleury and R. Brown, unpublished results.

[81] R. Kettner, J. Tittel, T. Basché, and C. Bräuchle, *J. Phys. Chem.* **98**, 6671, (1994).

[82] H. Z. Cummins and E. R. Pike (Eds.) *Photon Correlation and Light Beating Spectroscopy*, Plenum, New York, 1974.

[83] P. J. van der Zaag, J. P. Galaup, and S. Völker, *Chem. Phys. Lett.* **166**, 263, (1990).

[84] G. Zumofen and J. Klafter, *Chem. Phys. Lett.* **219**, 303, (1994).

[85] T. Holstein, S. K. Lyo, and R. Orbach, in *Laser Spectroscopy of Solids*, (Eds. W. M. Yen and P. M. Seltzer), Springer, Berlin, p. 39, where a similar analysis is given for energy transfer.

[86] Y. Kagan, *J. Low Temp. Phys.* **87**, 525, (1992).

[87] J. L. Skinner and H. P. Trommsdorff, *J. Chem. Phys.* **89**, 897, (1988).

[88] R. Rigler, J. Widengren, and U. Mets, in *Fluorescence Spectroscopy*, (Ed. O. Wolfbeis), Springer, Berlin, 1992, p. 13.

1.5 Theoretical Models for the Spectral Dynamics of Individual Molecules in Solids

J. L. Skinner

1.5.1 Introduction

Optical spectroscopy of dilute chromophores has proven to provide a very useful probe of the structural and dynamical properties of both crystalline and amorphous solids. This is because the transition frequency for a vibronic transition of a chromophore is generally very sensitive to the positions of the nearby atoms, ions, or molecules of the solid.

A typical absorption experiment involves a very large number of individual chromophore molecules, and the absorption line shape in a low-temperature solid is usually inhomogeneously broadened. This means that the line shape simply reflects the distribution of possible transition frequencies for the many chromophores, which is due to a distribution of local environments. Thus in this case it is clear that one can learn something about the structure of the solid from an analysis of the inhomogeneous lineshape [1–3].

For inhomogeneously broadened line shapes it necessarily follows that no information about time-dependent fluctuations of the chromophore's transition frequency (which I will call spectral dynamics) can be obtained from the line shape itself. This does not mean that such dynamic fluctuations do not occur; it simply means that either their amplitude is much smaller than the inhomogeneous line width or that their time scale is much longer than the inverse of the inhomogeneous line width. In either case these dynamic fluctuations are of great interest because they result from time-dependent changes in the local environments of chromophores, and hence can provide information about solid-state dynamics.

The experimental techniques of fluorescence line narrowing and hole burning were invented, in part, to access this dynamic information. They each involve selective excitation by a narrow-band laser of a nearly resonant subset of chromophores. The resulting fluorescence line shape or hole shape reflects the spectral dynamics of the members of this subset, unobscured by the other chromophores. In a similar vein, in the time-domain photon echo experiment, after the application of a short pulse the inhomogeneous dephasing of all of the chromophores is then rephased by a second pulse, and so the echo decay again reflects only transition frequency fluctuations.

Consider first the case of (substitutional) chromophores in crystals. The above techniques have all been used to measure the chromophores' spectral dynamics, which in this case are due to phonons. This situation is particularly simple because each chromophore molecule is influenced by the same kinds of phonons. Thus one says that these experiments measure the "homogeneous" line shape, implying that the inherent line shape of each chromophore due to its fluctuating transition frequency is the same. Furthermore, since the time scale for the fluctuations is much faster than the echo decay time (or the inverse of the hole or line width), the homogeneous line shape is always a motionally-narrowed Lorentzian [4].

The situation of chromophores in glasses, which are much more disordered than crystals, is more complicated for two reasons. First of all, the mechanism for the chromophores' fluctuations is thought to involve two-level systems (TLSs) [5, 6] rather than phonons. These TLSs are presumably distributed randomly throughout the glass, and it follows that different chromophore molecules will have statistically different spectral dynamics. Therefore it does not necessarily make sense to talk about a "homogeneous" line shape for a chromophore in a glass, since the dynamics probed by each chromophore is inherently different. Secondly, there appears to be a very wide distribution of time scales for TLS dynamics in glasses, extending from ps to beyond ks. The most important consequence of this is that experiments that involve different experimental time scales, like photon echoes and hole burning, will give different results [7].

Suppose that one could measure the absorption line shape of a single chromophore molecule. According to the above discussion, in a crystal each chromophore molecule is expected to have an identical (except for its center frequency) and reproducible Lorentzian "homogeneous" absorption line shape, reflecting the very fast phonon-induced fluctuations. In glasses, each molecule has a different environment of TLSs, and so each molecule should have a different line shape. By measuring the line shapes of many individual molecules one could access the distribution of spectral dynamical behaviors. Such a study could, in principle, provide more information than hole burning or echo experiments, which average over the spectral dynamics of many chromophores. Note, however, that the latter experiments, because of the much larger range of accessible experimental time scales, provide information that is complementary to that obtainable from single molecule studies.

In fact, single molecule spectroscopy (SMS) experiments have recently become a reality. The first experiments were performed on pentacene (the chromophore) in a *p*-terphenyl crystal [8–10]. I will focus here on the experiments of Ambrose, Basché, and Moerner [9, 10], which involved repeated fluorescence excitation spectrum scans of the same chromophore. For each chromophore molecule they found an identical (except for its center frequency) Lorentzian line shape whose line width is determined by fast phonon-induced fluctuations (and by the excited state lifetime), as discussed above. However, for each of a number of different chromophore molecules Moerner and coworkers found that the chromophore's center frequency changed from scan to scan, reflecting spectral dynamics on the time scale of many seconds! The transition frequencies of each of the chromophores seemed to sample a nearly infinite number of possible values. Plotting the transition frequency as a function of time produces what has been called a "spectral diffusion trajectory" (although the frequency fluctuations are not necessarily "diffusive"). These fascinating and totally unexpected results signaled a new mechanism for spectral dynamics in crystals, which must result from very slow molecular dynamics. This behavior has been interpreted as arising from a set of special TLSs in *p*-terphenyl crystal [11, 12]. Let me reiterate that because of their slow time scale these dynamics are not manifest in the usual (many-molecule) inhomogeneous absorption spectrum. More recent experiments on the chromophore terrylene (Tr) in hexadecane crystal (a Shpol'skii matrix) showed similar behavior [13].

Similar spectral dynamics of individual chromophores from repeated fluorescence excitation scans have subsequently been seen in amorphous hosts, for the systems Tr in polyethylene (PE) [14] and tetra-*t*-butyl-terrylene (TBT) in polyisobutylene (PIB) [15, 16]. In at least one instance [16] the chromophore samples far fewer frequencies than in the case of pentacene in *p*-terphenyl. The spectral diffusion trajectories are assumed to result from the flipping of those TLSs whose dynamics is slower than the scan time.

As mentioned earlier, the study of fluorescence excitation line shapes themselves can also provide a useful probe of the spectral dynamics of individual molecules in glasses. Unlike the case of pentacene in *p*-terphenyl crystal, where all molecules have the same line shape, individual chromophores in glasses do indeed have a variety of line shapes. This has been seen for perylene (Py) in PE [17], Tr in PE [18], Tr in polyvinylbutyral (PVB), polymethylmethacrylate (PMMA), and polystyrene (PS) [19], and for TBT in PE and PIB [15, 16]. The spectral broadening presumably is due to the flipping of those TLSs with dynamics faster than the scan time. There is a distribution of line shapes because, as noted earlier, different chromophore molecules are coupled to different sets of TLSs.

A third method for measuring spectral dynamics of individual molecules in glasses involves fluorescence intensity fluctuations during steady-state excitation at a fixed frequency. [20] This method has been applied to the systems of Tr in PE [18, 20] and TBT in PIB [15, 16]. In these experiments the fluorescence intensity fluctuates as the chromophore moves in and out of resonance because of coupling to flipping TLSs. Thus this very clever technique can provide a direct probe of TLS dynamics on all time scales.

It should be clear from the above that SMS presents a terrific opportunity to probe dynamics in both crystalline and amorphous solids at low temperatures. In order to provide a microscopic understanding of spectral dynamics and to analyze experimental results one needs theoretical models. The spectral dynamics in all of the experiments discussed above is assumed to arise from the coupling of the chromophore to one or more TLSs. In this chapter I will discuss the TLS model, and will attempt to provide a unified theoretical framework within which both the crystal and glass results, involving all three different experimental techniques, can be understood.

The organization of this chapter is as follows. Section 1.5.2 introduces the TLS model and discusses the dynamics of a single TLS. In Section 1.5.3 general results for the spectral diffusion kernel and the absorption line shape when the chromophore is coupled to an arbitrary number of TLSs are presented. Section 1.5.4 will discuss how the experimental observables (single molecule line shapes, spectral diffusion trajectories, and fluorescence intensity fluctuations) can be described with these results. In Section 1.5.5 attempts to analyze existing experiments will be described.

1.5.2 Dynamics of a single two-level system

In both the crystal and glass experiments the interesting spectral dynamics is thought to arise from the interaction of a chromophore with one or more TLSs that flip back and forth between their two states. In glasses the microscopic origin of TLSs is poorly understood, but they are assumed to be due to the motion of a collective coordinate involving a group of atoms in an asymmetric double well potential [21, 22]. In *p*-terphenyl crystal the microscopic origin of the relevant TLSs (those coupled to the chromophore) has been tentatively identified [11, 12], and the resulting picture is in accord with the above description.

This section considers a single asymmetric double-well potential. At low temperatures a quantum mechanical description is necessary, and only the lowest energy eigenstates will be relevant. If the energy asymmetry of the wells is not too great then it will be sufficient to describe the problem in terms of a two-state basis, where the two states are localized in each of the two wells of the potential. These two states compose the TLS.

Next let me discuss the mechanism for making transitions between these two states. The two states are of course coupled by a tunneling matrix element, but this coupling alone cannot produce "incoherent" population transfer between the states. So, in addition, coupling to phonons that is diagonal in the TLS basis is introduced. In the simplest model this coupling is linear in the phonon coordinates. There have been three approaches to determining the transition rate constants for this model, which involve treating: i) the tunneling exactly and the phonon coupling perturbatively [23]; ii) the phonon coupling exactly and the tunneling perturbatively [24]; iii) both the tunneling and phonon coupling perturbatively [25]. If both the tunneling and the phonon coupling are in fact small perturbations, then of course all three approaches give the same answer. If the energy splitting between the two states is $\Delta E \equiv \hbar\omega$, then in this case this model gives

$$k_u = k_0 n(\omega), \tag{1}$$

$$k_d = k_0[n(\omega) + 1], \tag{2}$$

where k_u and k_d are the rate constants for making "up" and "down" transitions respectively, k_0 is proportional to both the square of the tunneling matrix element and the square of the phonon coupling constant, and $n(\omega) = [\exp(\hbar\omega/kT) - 1]^{-1}$. This mechanism is called (one-)phonon-assisted tunneling. Other possible mechanisms exist, including two-phonon-assisted tunneling (via either linear [25] or quadratic [25, 26] phonon coupling), activated tunneling [27] (where the system climbs the vibrational ladder in one well before tunneling), or classical activated barrier crossing. The rate constants for each of these mechanisms have characteristic temperature dependences.

Whatever the mechanism, the simple first-order rate laws for the populations of the ground and excited TLS states, $P_0(t)$ and $P_1(t)$ respectively, are

$$\dot{P}_0(t) = -k_u P_0(t) + k_d P_1(t) \tag{3}$$

$$\dot{P}_1(t) = k_u P_0(t) - k_d P_1(t) \tag{4}$$

In thermal equilibrium the populations are

$$P_1 \equiv p = [e^{\Delta E/kT} + 1]^{-1} \tag{5}$$

$$P_0 = 1 - p = [e^{-\Delta E/kT} + 1]^{-1} \tag{6}$$

In order that the solutions to the above rate equations reach thermal equilibrium in the limit of long times the principle of detailed balance must be satisfied, which implies that

$$\frac{k_u}{k_d} = \frac{p}{1-p} = e^{-\Delta E/kT} \tag{7}$$

Let me define $P_{\alpha\beta}(t)$ to be the conditional probability that the TLS will be in state β at time t given that it was in state α at time 0 (α, $\beta = 0$, 1). These conditional probabilities are simply solutions to the rate equations subject to the appropriate initial conditions. They are given by

$$P_{\alpha\beta}(t) = \delta_{\alpha\beta} e^{-Kt} + P_\beta(1 - e^{-Kt}), \tag{8}$$

where $K = k_u + k_d$.

For what follows it will be convenient to define a stochastic occupation variable ξ for the TLS, such that when the system is in its ground state $\xi = 0$ and when it is in its excited state $\xi = 1$. The average value of any function of ξ is given by

$$\langle f(\xi) \rangle = \sum_\xi P_\xi f(\xi). \tag{9}$$

Thus for example $\langle \xi \rangle = p$. The conditional probabilities allow one to calculate the statistical average of any pair of functions of ξ at two different times:

$$\langle f(\xi(t))g(\xi(0)) \rangle = \sum_{\xi^0} \sum_\xi P_{\xi^0} P_{\xi^0 \xi}(t) g(\xi^0) f(\xi). \tag{10}$$

For example, the time correlation function of ξ is

$$\langle \xi(t)\xi(0) \rangle = p(1-p)e^{-Kt} + p^2. \tag{11}$$

1.5.3 Spectral dynamics of a chromophore coupled to one or many two-level systems

As discussed earlier, the transition frequency for a particular vibronic transition of a chromophore is sensitive to its surroundings. The transition frequency fluctuations can be assumed to be solely due to TLSs nearby to the chromophore. Therefore, the transition frequency of an individual chromophore will be the sum of a time-independent term involving any non-TLS interactions, and a fluctuating term that depends on the instantaneous configurations of nearby TLSs. Thus the (angular) frequency at time t is

$$\omega(t) = \bar{\omega} + \sum_j \xi_j(t) v_j, \tag{12}$$

where the sum is over those TLSs interacting with the chromophore, $\bar{\omega}$ is the chromophore's transition frequency when all of these TLSs are in their ground states, $\xi_j(t)$ is the time-dependent occupation variable for the jth TLS, and v_j is the perturbation of the transition frequency when TLS j is excited. v_j depends on the distance r_j of the chromophore from TLS j, and will typically be taken to result from dipolar interactions; hence it decreases as $1/r_j^3$. Each TLS j is, in general, characterized by its own set of static and dynamic parameters p_j and K_j.

Since each of the $\xi_j(t)$ is a stochastic variable, it follows that $\omega(t)$ is also. Each of the SMS experimental observables to be described in Section 1.5.4 measures statistical properties of this fluctuating transition frequency. In this section I will discuss these properties.

The simplest way to characterize a stochastic variable is with its correlation function, defined in this case by

$$C(t) = \langle \omega(t)\omega(0) \rangle - \langle \omega \rangle^2. \tag{13}$$

Since the $\xi_j(t)$ are assumed to be uncorrelated with each other it is easy to show from Eq. (11) that

$$C(t) = \sum_j p_j(1 - p_j) v_j^2 e^{-K_j t}. \tag{14}$$

For more complicated averages it will be useful to proceed more generally. For a Markovian process such as the one we are considering, the two fundamental statistical quantities describing the fluctuating frequency are the singlet and joint probabilities $P(\omega)$ and $P(\omega, t; \omega_0, 0)$. Thus $P(\omega)$ is the equilibrium distribution of frequencies visited by the chromophore, and $P(\omega, t; \omega_0, 0)$ is the probability distribution that the chromophore has frequency ω at time t and had frequency ω_0 at time 0. For the special case that all TLSs have the same p_j and K_j, exact expressions for these quantities have recently been derived [28]. Generalization by allowing for different p_j and K_j leads to

$$P(\omega) = \int_{-\infty}^{\infty} \frac{\mathrm{d}\tau}{2\pi} e^{i(\omega-\bar{\omega})\tau} \prod_j [1 + p_j(e^{-iv_j\tau} - 1)], \tag{15}$$

$$P(\omega, t; \omega_0, 0) = \int_{-\infty}^{\infty} \frac{\mathrm{d}\tau_1}{2\pi} \int_{-\infty}^{\infty} \frac{\mathrm{d}\tau_2}{2\pi} e^{i(\omega_0-\bar{\omega})\tau_1} e^{i(\omega-\bar{\omega})\tau_2} \prod_j [1 + p_j(e^{-iv_j(\tau_1+\tau_2)} - 1)$$

$$+ p_j(1-p_j)(e^{-K_jt} - 1)(e^{-i\tau_1 v_j} - 1)(e^{-i\tau_2 v_j} - 1)]. \tag{16}$$

With these quantities one can write the average of a function of ω by

$$\langle f(\omega) \rangle = \int_{-\infty}^{\infty} \mathrm{d}\omega f(\omega) P(\omega), \tag{17}$$

and the average of two different functions at different times by

$$\langle f(\omega(t))g(\omega(0)) \rangle = \int_{-\infty}^{\infty} \mathrm{d}\omega \int_{-\infty}^{\infty} \mathrm{d}\omega_0 P(\omega, t; \omega_0, 0) f(\omega)g(\omega_0). \tag{18}$$

A related quantity, the conditional probability density (also called the spectral diffusion kernel) $P(\omega, t|\omega_0)$ is defined by

$$P(\omega, t|\omega_0) = P(\omega, t; \omega_0, 0)/P(\omega_0). \tag{19}$$

$P(\omega, t|\omega_0)$ is the probability density that the chromophore has transition frequency ω at time t given that it had frequency ω_0 at time 0. While the functional form of this spectral diffusion kernel is quite complicated in general, at short times certain simplifications occur. In particular, if: the positions of the TLSs occupy a regular lattice in three-dimensional space, all of the relaxation rates K_j are the same, all of the occupation probabilities p_j are the same and equal to $1/2$ (the high-temperature limit), and the perturbations v_j are dipolar, then it was shown by Klauder and Anderson [29] and more recently by Zumofen and Klafter [30] that the spectral diffusion kernel is Lorentzian:

$$P(\omega, t|\omega_0) = \frac{\Gamma(t)/\pi}{(\omega - \omega_0)^2 + \Gamma(t)^2}, \tag{20}$$

with a time-dependent width, $\Gamma(t)$, that is linear in t. This result was recently generalized to the case where the p_j are the same but not necessarily equal to $1/2$ [28]. Other results for more general power law interactions and other spatial dimensions are given by Zumofen and Klafter [30].

A less detailed but still very informative characterization of the fluctuating frequency is the distribution of spectral jumps, defined to be $P(\Delta; t)$, which is the probability density that the transition frequency will change by an amount Δ in time t. It is given by [28]

$$P(\Delta; t) = \int_{-\infty}^{\infty} \frac{d\tau}{2\pi} e^{i\Delta\tau} \prod_j [1 + 2p_j(1 - p_j)(1 - e^{-K_j t})(\cos(v_j\tau) - 1)]. \qquad (21)$$

Another important quantity that depends on $\omega(t)$ is the spectral line shape function, defined by [31]

$$I(\omega) = Re\left\{ \frac{1}{\pi} \int_0^{\infty} dt\, e^{i\omega t} e^{-t/2T_1} \phi(t) \right\}, \qquad (22)$$

where

$$\phi(t) = \left\langle \exp\left(-i \int_0^t d\omega(\tau) \right) \right\rangle. \qquad (23)$$

T_1 is the excited state lifetime. Unlike the spectral diffusion kernel or frequency correlation function, which involve only two different times, $\phi(t)$ depends on $\omega(\tau)$ at all times between 0 and t. The seemingly formidable problem of calculating $\phi(t)$ has been solved in general by Kubo and Anderson [31–34], and the solution for the current model is [35],

$$\phi(t) = e^{-i\bar{\omega}t} \prod_j \phi_j(t), \qquad (24)$$

$$\phi_j(t) = e^{-(\alpha_j + ip_j v_j)t} \left[\cosh(\Omega_j t) + \frac{\alpha_j}{\Omega_j} \sinh(\Omega_j t) \right], \qquad (25)$$

$$\Omega_j = \sqrt{\frac{K_j^2}{4} - \frac{v_j^2}{4} - i\left(p_j - \frac{1}{2}\right)v_j K_j}, \qquad (26)$$

$$\alpha_j = \frac{K_j}{2} - i\left(p_j - \frac{1}{2}\right)v_j. \qquad (27)$$

1.5.4 Experimental observables

The most straightforward experimental measurement of the spectral dynamics of a single molecule simply involves the absorption (or fluorescence excitation) line shape. Within the TLS model described herein one sees from Eqs. (22) and (23) that the line shape does indeed depend on the fluctuating frequency $\omega(t)$. While this lineshape formula is usually used to calculate the spectrum for an ensemble of chromophores, in principle it is also appropriate for individual chromophores. To measure the spectrum of a single molecule one must of course collect many photons, which means that the molecule must be cycled between its ground and excited state many times. Each time the molecule reappears in its ground state in effect starts a new experiment. In this manner even the line shape for a single molecule represents an ensemble average over the stochastic process. However, an implicit assumption in this argument is that the time it takes to perform the measurement, τ, is long compared to all TLS relaxation times, $1/K_j$. Alas, this is not usually the case for single molecule experiments in glasses, since τ is on the order of seconds and $1/K_j$ can be much longer.

In this more general case, the theory of single molecule line shapes has not yet been fully developed. Nonetheless, some understanding can be gained by arguing along the following lines, adapted from the work by Fleury et al. [18], but reformulated within the present theoretical framework. Let us first divide the TLSs into two classes, those with $K_j > 1/\tau$, which we will call fast, and those with $K_j < 1/\tau$, which we will call slow. It must be true that if $K_j \ll 1/\tau$ then the dynamics of these very slow TLSs simply would not be relevant to the spectrum, and the (static on the experimental time scale) configurations of these TLSs would only lead to a renormalization of the chromophore's frequency $\bar{\omega}$, which is of no interest. So as a first approximation let us simply neglect all the slow TLSs. If for all the fast TLSs $K_j \gg 1/\tau$, then τ would be long compared to the relevant TLS relaxation times, and the line shape would be given by Eq. (22), but the product in Eq. (24) is over only the fast TLSs. While it is not strictly true that $K_j \gg 1/\tau$ for all fast TLSs, we will nevertheless take this approach.

Now let us further divide the fast TLSs into two groups, those with $K_j > v_j$ and those with $K_j < v_j$. TLSs with $K_j \gg v_j$ would lead to "motional narrowing" [31], while TLSs with $K_j \ll v_j$ would lead to "inhomogeneous broadening" (in that the line shape would simply reflect the distribution of frequencies sampled by the chromophore). With this in mind we label the TLSs for which $K_j > v_j$ by m instead of j, and those for which $K_j < v_j$ by i instead of j. Evaluating Eq. 25 for $\phi_j(t)$ in the limit $K_j \gg v_j$ leads to (replacing j by m) [35]

$$\phi_m(t) = e^{-ip_m v_m t} e^{-[p_m(1-p_m)v_m^2/K_m]t}. \tag{28}$$

Evaluating Eq. (25) in the limit $K_j \ll v_j$ gives (replacing j by i) [35]

$$\phi_i(t) = 1 + p_i(e^{-iv_i t} - 1). \tag{29}$$

As a first approximation one might simply neglect the m TLSs [18], and use the above expression for $\phi_i(t)$ for all the i TLSs. The spectrum is then given approximately by

$$I(\omega) = \text{Re}\left\{ \frac{1}{\pi} \int_0^\infty dt\, e^{i(\omega-\bar{\omega})t} e^{-t/2T_1} \prod_i [1 + p_i(e^{-iv_i t} - 1)] \right\}, \tag{30}$$

which looks like an inhomogeneously broadened line shape due to the TLSs labeled i [3], with an additional "dephasing" contribution from the excited state lifetime.

Neglecting for a moment the lifetime contribution, the line shape can be characterized by its variance

$$\langle \omega^2 \rangle_i - \langle \omega \rangle_i^2 = \sum_i p_i (1 - p_i) v_i^2, \tag{31}$$

where the subscript i indicates that only the stochastic variables ξ_i are included in the average. Since each individual chromophore has a distinct set of i TLSs in its environment, producing different perturbations v_i, each chromophore's lineshape would have a different variance. The distribution of variances provides a way to characterize the distribution of single molecule line shapes, both experimentally and theoretically.

In the second type of experiment that measures single molecule spectral dynamics one performs repeated fluorescence excitation scans of the same molecule. In each scan the line shape is described as above, but now there is the possibility that the center frequency of the line will change from scan to scan because of slow fluctuations. Thus one can measure the center frequency as a function of time, producing what has been called a spectral diffusion trajectory. This trajectory can, in principle, be characterized completely by the spectral diffusion kernel of Eqs. (16) and (19), but of course it must be understood that only the slow ($K_j < 1/\tau$) TLSs contribute. In fact, the experimental trajectories are really too short to be analyzed with this spectral diffusion kernel. Instead, it is useful [11, 12] to consider three simpler characterizations of the spectral diffusion trajectories: the frequency-frequency correlation function in Eq. (14), the distribution of frequencies from Eq. (15), and the distribution of spectral jumps from Eq. (21). For this application of the theoretical results, in all three of these formulas j should be replaced by s, the labels for the slow TLSs.

The third type of experimental observable for single molecule spectral dynamics is the correlation of fluorescence intensity fluctuations. In this experiment one pumps the chromophore under steady-state conditions at a fixed frequency ω_L and measures the time-dependent intensity $I(t)$, whose fluctuations are analyzed by the correlation function

$$C_I(t) = \langle I(t)I(0) \rangle - \langle I \rangle^2. \tag{32}$$

A rough attempt to understand these experiments proceeds similarly to the discussion of the line shape. Considering only the TLSs with $K_j < v_j$, and labeling them again by i (although in this case there is no lower limit to the value of K_j), their effect will be to modulate the center frequency of the line shape. Thus we write

$$I(t) = \frac{\Gamma/\pi}{\left(\omega_L - \omega(t) \right)^2 + \Gamma^2}, \tag{33}$$

where $\omega(t)$ is given by Eq. (12) but with j replaced by i, and $\Gamma = 1/2T_1$.

The intensity correlation function depends upon $\langle I \rangle$ and $\langle I(t)I(0) \rangle$, both of which can be calculated from Eqs. (15–18) (but j must be replaced by i). Performing the frequency integrals therein gives

$$\langle I \rangle = \int_{-\infty}^{\infty} \frac{d\tau}{2\pi} e^{i(\omega_L - \bar{\omega})\tau} e^{-\Gamma|\tau|} \prod_i [1 + p_i(e^{-iv_i\tau} - 1)], \tag{34}$$

$$\langle I(t)I(0) \rangle = \int_{-\infty}^{\infty} \frac{d\tau_1}{2\pi} \int_{-\infty}^{\infty} \frac{d\tau_2}{2\pi} e^{i(\omega_L - \bar{\omega})\tau_1} e^{i(\omega_L - \bar{\omega})\tau_2} e^{-\Gamma|\tau_1|} e^{-\Gamma|\tau_2|}$$

$$\times \prod_i [1 + p_i(e^{-iv_i(\tau_1 + \tau_2)} - 1)$$

$$+ p_i(1 - p_i)(e^{-K_i t} - 1)(e^{-i\tau_1 v_i} - 1)(e^{-i\tau_2 v_i} - 1)]. \tag{35}$$

If the chromophore is coupled to only one i TLS, the situation simplifies considerably, and from the above one finds that

$$C_I(t) = (I_1 - I_0)^2 p(1 - p)e^{-Kt}, \tag{36}$$

where $I_0(I_1)$ is the intensity when the TLS is in its ground (excited) state. I_0 and I_1 are given by

$$I_o = \frac{\Gamma/\pi}{(\omega_L - \bar{\omega})^2 + \Gamma^2}, \tag{37}$$

$$I_1 = \frac{\Gamma/\pi}{(\omega_L - \bar{\omega} - v)^2 + \Gamma^2}. \tag{38}$$

An equivalent result was derived earlier by Fleury et al. [18]. Perhaps an easier way to obtain Eq. (36) is to write

$$I(t) = I_0 + (I_1 - I_0)\xi(t). \tag{39}$$

The desired formula then follows immediately from the results in Section 1.5.2.

1.5.5 Analysis of experiments

Let me begin by discussing the SMS experiments involving chromophores in crystals, specifically pentacene in p-terphenyl [9, 10]. The fluorescence excitation line shapes of individual pentacene molecules are all identical, and the line widths have the lifetime-limited value at the lowest temperatures, and increase with temperature due to pure dephasing by phonons. Thus in this case the chromophores are not coupled to any fast ($K_j > 1/\tau$) TLSs, and there is nothing particularly interesting about the single molecule line shapes. What is very interesting, however, is that for some molecules the center frequency of the line shape changes from scan to scan, which is what we have called spectral diffusion. Within the present framework this implies that the chromophores are coupled to TLSs of the slow variety. In addition, the chromphore visits a very large number of possible frequencies, implying that the chromophore is coupled to a large number of TLSs.

Our physical picture of the situation is as follows [12]. The low-temperature phase of *p*-terphenyl crystal is known to have two crystallographically distinct degenerate ground states, corresponding to different orientational ordering of the central phenyl rings [36]. A real crystal will therefore have domains of these two ground states, separated by domain walls. In each of the domains the potential curve for reorientation of a single central phenyl ring is very asymmetric, and at the temperatures of the experiment reorientation does not occur. However, at a perfect domain wall this potential is symmetric, thus allowing for the possibility of central phenyl ring flips. The left and right states of this symmetric potential correspond to the two states of the TLS. The rate of TLS flipping is expected to be very slow since this process involves tunneling of carbon through a substantial barrier. Chromophores that happen to be near a domain wall sense a large number of flipping TLSs and therefore show spectral diffusion behavior.

As discussed above, the simplest characterization of a spectral diffusion trajectory involves the transition frequency correlation function $C(t)$ from Eq. (14) (but with *j* replaced by *s*). The experimental results [9, 10] for a given chromophore can be fit with an exponentially decaying $C(t)$, suggesting that all of the TLSs might have the same decay rate K [12]. If we also assume that each TLS has the same energy splitting, analysis of the temperature dependence of $C(t)$ shows that this energy splitting is small (on the order of 10 K) but nonzero, and varies from domain wall to domain wall [12]. The origin of this energy splitting is not precisely understood. Possibilities include anything that breaks the symmetry of a perfect planar domain wall, including thermal fluctuations of the interface, the presence of other defects, or indeed, the chromophore itself.

Analysis [12] of the temperature dependences of the frequency distribution $P(\omega)$ and of the jump distribution $P(\Delta; t)$ for individual chromophores provides what seems to be compelling confirmation of the above physical picture. (In the actual analysis of the data the expressions for $P(\omega)$ and $P(\Delta; t)$ in Eqs. (15) and (21) were modified to include the effect of additional experimental noise [12].) For one chromophore the temperature dependence of the TLS flip rate is consistent with one-phonon-assisted tunneling, but for another it is not. More experimental studies on more molecules, for longer times, and over a wider range of temperatures, would help determine the mechanisms of TLS flipping, and would also provide further overall evidence to support (or refute) our picture.

In glasses, single molecule line shapes show significant variation from molecule to molecule, and typically the line width is larger than the lifetime-limited value. This implies both that the chromophores are coupled to fast TLSs, and that different chromophores are coupled to sets of TLSs with different parameters. Therefore these experiments can, in principle, provide a wealth of information about TLS dynamics in glasses. The only attempt at theoretical analysis of these line shapes was performed by Fleury et al. [18]. Their analysis was along the lines of the discussion in

Section 1.5.4. In the high-temperature limit $p_i \approx 1/2$ in Eq. (32), and so the variance of the distribution of frequencies due to the i TLSs is $\langle \omega^2 \rangle_i - \langle \omega \rangle_i^2 = (1/4) \sum_i v_i^2$. Assuming that this variance of the distribution is proportional to the FWHM of the distribution, they then incorporated the excited state lifetime contribution to arrive at the line width for a single molecule. Assuming that the chromophore–TLS interactions are dipolar and that the TLSs are randomly distributed in space then leads to a distribution of variances (for different chromophores), which can be calculated with the statistical theory of Stoneham [1]. This approach produces a distribution of line widths that is in good agreement with experiments on Tr in PE [18]. The experimental line widths [18] showed a variety of temperature dependences, but this could not be analyzed with their approach because of the implicit high-temperature assumption. Line width distributions have been measured for a number of other systems [15–17, 19], but no similar theoretical analysis has been performed. One notable result, however, is that for some systems [19] the average of the single molecule line widths is approximately equal to one half of the hole width, as measured by hole burning experiments. This confirms the idea that hole burning experiments, which involve an ensemble of chromophores, measure the average single-molecule line shape.

Spectral diffusion trajectories due to spontaneous (rather than light-induced) fluctuations have been measured for Tr in PE [14] and for TBT in PIB [15, 16]. As in the crystalline case these trajectories reflect dynamics of the slow TLSs. The three published trajectories show that in two cases the chromophore visits a large number of frequencies, and in one case, only four. In this latter case the chromophore is presumably strongly coupled to two TLSs. A correlation function analysis was applied to the PIB system, but for neither the PIB nor the PE system was a temperature-dependent study reported.

Fluorescence intensity fluctuations were first measured for Tr in PE by Zumbusch et al. [20]. Several of the correlation function decays were distinctly nonexponential, presumably indicating that in these cases the chromophore was coupled to at least two TLSs. However, in some cases the decay was exponential, and the results were analyzed with Eq. (36), which assumes that the chromophore is coupled to a single TLS. In this manner the relaxation rate K was obtained. Furthermore, the temperature dependence of K was also determined. In several cases it was consistent with one-phonon-assisted tunneling, but in other cases it was not. Fluorescence intensity fluctuations have also been measured for TBT in PIB [16]. At low laser powers the decays were single exponential, although the corresponding spectral diffusion trajectories showed that many different frequencies were sampled. Tittel et al. argued that in this case a single TLS is strongly coupled to the chromophore and therefore dominates the intensity fluctuations, while weaker coupling to a number of other TLSs contributes to the spectral diffusion trajectory [16].

1.5.6 Conclusion

We are now probably entering a very fruitful period as far as using single-molecule spectroscopy to probe molecular dynamics in solids is concerned. The spectral diffusion experiments on pentacene in *p*-terphenyl crystals were ground-breaking in that a new phenomenon was discovered. For this system we have learned a significant amount about molecular dynamics at domain walls and how that is probed by a chromophore. This system deserves further study because in this case we are able to identify the nature of the TLSs that appear to be responsible for spectral diffusion, which makes it easier to perform theoretical analysis and consequently to develop a full understanding. In this sense this system has served and will continue to serve as a very useful prototype for more disordered amorphous systems.

Extremely exciting experimental data for glasses are now beginning to emerge. It has been shown that line shape measurements, fluorescence intensity fluctuations, and spectral diffusion trajectories can all be used to probe TLS dynamics on different time scales. Furthermore, as has been emphasized already, these experiments on individual molecules will provide information complementary to that obtained from more traditional echo and hole burning experiments. At this point what we need is more data. In an ideal world all three of the above experiments would be performed on the same individual molecule at a variety of temperatures, and then would be repeated on many molecules, and all of the above would be repeated for several different systems. Although the basic theoretical apparatus is in place for analyzing these experiments, more refined theoretical results will surely be needed.

Acknowledgments

I am grateful for support from the National Science Foundation (Grant No. CHE95-26815) and from the American Chemical Society (Grant No. PRF28464-AC6). I thank Dr. Eitan Geva for insightful conversations.

References

[1] A. M. Stoneham, *Rev. Mod. Phys.* **41**, 82, (1969).
[2] H. M. Sevian and J. L. Skinner, *Theoretica Chimica Acta* **82**, 29, (1992).
[3] D. L. Orth, R. J. Mashl, and J. L. Skinner, *J. Phys.-Cond. Matter* **5**, 2533, (1993).
[4] J. L. Skinner, *Annu. Rev. Phys. Chem.* **39**, 463, (1988).
[5] P. W. Anderson, B. I. Halperin, and C. M. Varma, *Phil. Mag.* **25**, 1, (1972).
[6] W. A. Phillips, *J. Low Temp. Phys.* **3/4**, 351, (1972).
[7] M. Berg, C. A. Walsh, L. R. Narasimhan, K. A. Littau, and M. D. Fayer, *Chem. Phys. Lett.* **139**, 66, (1987).
[8] M. Orrit and J. Bernard, *Phys. Rev. Lett.* **65**, 2716, (1990).
[9] W. P. Ambrose and W. E. Moerner, *Nature*, **349**, 225, (1991).
[10] W. P. Ambrose, Th. Basché, and W. E. Moerner, *J. Chem. Phys.* **95**, 7150, (1991).
[11] P. D. Reilly and J. L. Skinner, *Phys. Rev. Lett.* **71**, 4257, (1993).
[12] P. D. Reilly and J. L. Skinner, *J. Chem. Phys.* **102**, 1540, (1995).
[13] W. E. Moerner, T. Plakhotnik, T. Irngartinger, M. Croci, V. Palm, and U. P. Wild, *J. Phys. Chem.* **98**, 7382, (1994).
[14] P. Tchénio, A. B. Myers, and W. E. Moerner, *J. Lumin.* **56**, 1, (1993).
[15] R. Kettner, J. Tittel, Th. Basché, and C. Bräuchle, *J. Phys. Chem.* **98**, 6671, (1994).
[16] J. Tittel, R. Kettner, Th. Basché, C. Bräuchle, H. Quante, and K. Müllen, *J. Lumin.* **64**, 1, (1995).

[17] Th. Basché, W. P. Ambrose, and W. E. Moerner, *J. Opt. Soc. Am.* **B9**, 829, (1992).
[18] L. Fleury, A. Zumbusch, M. Orrit, R. Brown, and J. Bernard, *J. Lumin.* **56**, 15, (1993).
[19] B. Kozankiewicz, J. Bernard, and M. Orrit, *J. Chem. Phys.* **101**, 9377, (1994).
[20] A. Zumbusch, L. Fleury, R. Brown, J. Bernard, and M. Orrit, *Phys. Rev. Lett.* **70**, 3584, (1993).
[21] A. Heuer and R. Silbey, *Phys. Rev. Lett.* **70**, 3911, (1993).
[22] D. Dab, A. Heuer, and R. J. Silbey, *J. Lumin.* **64**, 95, (1995).
[23] J. L. Skinner and H. P. Trommsdorff, *J. Chem. Phys.* **89**, 897, (1988).
[24] B. Jackson and R. Silbey, *J. Chem. Phys.* **78**, 4193, (1983).
[25] T. Holstein, S. K. Lyo, and R. Orbach, In: *Laser Spectroscopy of Solids* (Ed. W. M. Yen and P. M. Selzen), Springer, Berlin, 1981.
[26] R. Silbey and H. P. Trommsdorff, *Chem. Phys. Lett.* **165**, 540, (1990).
[27] P. E. Parris and R. Silbey, *J. Chem. Phys.* **83**, 5619, (1985).
[28] P. D. Reilly and J. L. Skinner, *J. Chem. Phys.* **101**, 965, (1994).
[29] J. R. Klauder and P. W. Anderson, *Phys. Rev.* **125**, 912, (1962).
[30] G. Zumofen and J. Klafter, *Chem. Phys. Lett.* **219**, 303, (1994).
[31] R. Kubo, *Adv. Chem. Phys.* **15**, 101, (1969).
[32] P. W. Anderson, *J. Phys. Soc. Jpn.* **9**, 316, (1954).
[33] R. Kubo, *J. Phys. Soc. Jpn.* **9**, 935, (1954).
[34] R. Kubo, In: *Fluctuation, Relaxation, and Resonance in Magnetic Systems*, (Ed. D. TerHaar), Oliver and Boyd, Edinburgh, 1962.
[35] P. D. Reilly and J. L. Skinner, *J. Chem. Phys.* **101**, 959, (1994).
[36] C. Ecolivet, B. Toudic, and M. Sanquer, *J. Chem. Phys.* **81**, 599, (1984).

1.6 Magnetic Resonance of Single Molecular Spins

J. Wrachtrup, C. von Borczyskowski, J. Köhler, J. Schmidt

1.6.1 Introduction

The development of electron paramagnetic resonance (EPR) and nuclear magnetic resonance (NMR) spectroscopy ranks among the most important advances in physics and chemistry of the last 50 years. Few other techniques offer such a direct and detailed insight into events at the atomic and nuclear level. A serious handicap in the application of these methods is that the sensitivity is fairly low owing to the fact that the energies of the microwave photons in the case of EPR and the radiofrequency photons in the case of NMR are relatively small and that consequently a large number of spins is needed to obtain a detectable absorption signal. For instance for NMR usually 10^{16} to 10^{18} spins are necessary and for EPR 10^{10} to 10^{12} spins.

In the course of time several methods have been devised to enhance the sensitivity of magnetic-resonance spectroscopy. The first seeks to increase the thermal population distribution over the spin levels [1]. The second method is based on the idea to transfer the detection of the magnetic-resonance absorption to the optical domain with a concomitant increase in photon energy [2]. This method carries the general name Optical Detection of Magnetic Resonance or ODMR. The first ODMR experiment on organic molecules in the condensed phase has been carried out in 1967 by Sharnoff [3] who observed the "$\Delta m = 2$" transition in the lowest triplet state of naphthalene as a change in the intensity of the phosphorescence, soon followed by the detection of the "$\Delta m = 1$" transitions in phosphorescent phenanthrene by Kwiram [4] and by Schmidt et al. [5] in phosphorescent quinoxaline. Alternatively it proved possible to detect the magnetic resonance signal in nonphosphorescent triplet-state molecules as changes in the fluorescence intensity (FDMR or fluorescence-detected magnetic resonance) [6]. In contrast to these two methods, where the optical emission is used as means of detection, the technique of Absorption Detected Magnetic Resonance (ADMR) relies on the fact that the absorption of a microwave photon in the metastable triplet state leads to a change in the absorption intensity. The attractive feature of ODMR spectroscopy is the high sensitivity. For instance in favorable cases as few as 10^6 spins can be detected [2]. Moreover, the optical excitation allows the selection of specific molecular species.

The development of optical single-molecule spectroscopy during the last couple of years has opened a completely new way to obtain information about individual molecules. The natural question which arises is whether it is possible to perform magnetic resonance experiments on individual, optically selected molecules using

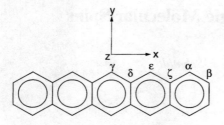

Figure 1. Molecular structure of pentacene ($C_{22}H_{14}$), molecular axis system and labeling of the inequivalent carbon positions.

ODMR-type methods. The answer is affirmative [7, 8] and in this contribution we discuss a number of results on the system pentacene doped into a *p*-terphenyl host crystal (see Fig. 1) which demonstrate that such experiments are feasible and that it is possible to extract information about a single molecular spin and its interaction with its environment and the resonant radiation field.

1.6.2 Principles of the experiment

The principles of fluorescence-detected, single-molecule spectroscopy (SMS) have already been discussed elsewhere in this book. Below 193 K the *p*-terphenyl crystal has four crystal sites, called P_1 to P_4 [9], which can be substitutionally occupied by the pentacene guest molecules. This results in four distinct spectral origins in the absorption spectrum of pentacene in *p*-terphenyl at positions $16\,883\,cm^{-1}$ (O_1), $16\,887\,cm^{-1}$ (O_2), $17\,006\,cm^{-1}$ (O_3), and $17\,065\,cm^{-1}$ (O_4) [10–13]. To avoid confusion we denote the *p*-terphenyl lattice sites by P_1 to P_4 and the pentacene spectral origins by O_1 to O_4. Owing to considerable differences in the photo-physical parameters [13] only pentacene molecules responsible for the O_1 and O_2 spectral origins, which have a high fluorescence yield (radiative lifetime 22 ns) are suitable for single-molecule spectroscopy. Moreover the magnetic properties of the lowest, metastable triplet state are well known from earlier FDMR and Electron Spin Echo experiments [14]. In Fig. 2 we show a schematic representation of the ground state and the lowest excited states in the so-called Jablonski diagram. A single pentacene molecule in the *p*-terphenyl host is selected at low temperature (1.2–1.5 K) by tuning a narrow-band laser to the absorption wavelength of the $^1S_0 \leftarrow {}^1S_1$ zero-phonon transition (for details see Section 1.1). Under influence of the laser field the molecule undergoes repeated excitation-emission cycles between $^1S_0$ and $^1S_1$. The quantum yield for intersystem crossing (ISC) to the lowest triplet state $^3T_1$ for the spectral sites O_1 and O_2 is approximately 0.5%. The decay from $^3T_1$ to the singlet ground state proceeds via a non-radiative process leading to a mean lifetime of about 50 μs.

The three sublevels of $^3T_1$ are split even in the absence of a magnetic field owing to the anisotropic dipolar interaction of the two unpaired electron spins. The sublevels are labelled $|X\rangle$, $|Y\rangle$, and $|Z\rangle$ and are related to the principal axis of the fine-structure tensor, which by symmetry are constrained to lie parallel to the three two-fold symmetry axes of pentacene. The zero-field splitting in the case of pentacene is in the order of 1.5 GHz [14].

Figure 2. Schematic representation of the ODMR effect on a single molecule. Part (a) of the figure shows the five-level scheme relevant for the experiment. k_{u2} denote the population rates of the sublevels $|U\rangle$ $(U = X, Y, Z)$ and k_{1u} represent the depopulation rates. The relative magnitudes of the rates with respect to each other are indicated by the thickness of the arrows. Part (b) depicts the temporal evolution of the photon electron pulses as they are created by the detection system. Upon irradiation with microwaves in resonance with the $|X\rangle-|Z\rangle$ transition the dark periods between the photon bunches are lengthened.

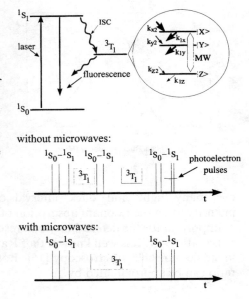

The populating, k_{u2} $(u = x, y, z)$, and depopulating probabilities, k_{1u} $(u = x, y, z)$, for the three sublevels differ significantly as is indicated in Fig. 2(a) by the relative size of the corresponding arrows. This is the result of the high selectivity of the ISC process which is related to the fact that the spin orbit coupling can only mix singlet character into specific triplet sublevels. It is known that for pentacene in p-terphenyl the $|X\rangle$ and $|Y\rangle$ sublevels have a much higher populating probability than $|Z\rangle$. These two levels also have a much shorter lifetime. In a conventional FDMR experiment one excites an ensemble of pentacene molecules with a continuous light source and creates a considerable population difference between the three triplet sublevels. Irradiation with microwaves resonant with either the $|X\rangle-|Z\rangle$ or $|Y\rangle-|Z\rangle$ transition leads to a redistribution of the population of the two levels involved in the resonance and hence to a change of the average lifetime of the triplet state. This in turn affects the population of the ground state and, since the light source excites the system permanently, this leads to a change of the fluorescence intensity.

In the case of a single molecule the same experiment can be carried out. When a single molecule resides in the triplet state the molecule does not emit light. However as soon as it has decayed to the ground state it is subjected to the laser excitation and it undergoes excitation-emission cycles between $^1S_0$ and $^1S_1$ until the ISC process takes it back to the triplet state. Consequently, the emission of light from a single molecule consists of light periods, during which photons are emitted, and dark periods when the molecule is in the $^3T_1$ state (see Fig. 2(b)). This phenomenon is called "photon-bunching" [15] and it is one of the signatures that the emission originates from a single molecule and not from an ensemble. Irradiation with microwaves resonant with the $|X\rangle-|Z\rangle$ or $|Y\rangle-|Z\rangle$ transition transfers the molecule from a short-lived to a long-lived sublevel thus lengthening the dark interval. When averaging

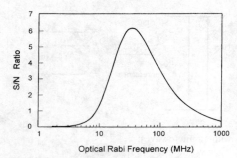

Figure 3. Signal-to-noise (S/N) ratio of the ODMR signal as a function of the square root of the laser power (proportional to the Rabi frequency of the optical transition). The calculation is performed using typical photophysical parameters (see Table 4) for the system pentacene in p-terphenyl (O_1, O_2 spectral sites). A detection efficiency of 0.1%, a C_0 of 0.06 and an integration time of 1 s has been assumed.

over many "light" and "dark" intervals one observes a reduction of the fluorescence intensity upon the resonant absorption of the microwave field.

Important for the detectability of a resonant microwave transition is the signal-to-noise ratio S/N. It is well known, that Rayleigh scattering is the limiting noise factor in single-molecule spectroscopy [16]. Both depend on the laser intensity. The S/N ratio can be approximated by

$$\frac{S}{N} = \frac{(I_0 - I_{MW})\Phi_D \tau}{\sqrt{I_0 \Phi_D \tau + C_0 P_0 \tau + N_d \tau}}, \tag{1}$$

Here Φ_D is the experimental detection efficiency for fluorescence photons and τ is the integration time. The relevant noise contributions are the photon noise of the laser light given by $I_0 \Phi_D$, the background photons proportional to the laser power P_0 and the dark noise of the detection system N_d. I_0 is the fluorescence intensity without and I_{MW} with microwave irradiation. Both values can be calculated from the known absorption cross section for the photon flux [17] and with the aid of Eq. (10), vide infra. C_0 is an empirically determined constant [16]. Making use of the intramolecular rate parameters [18] the S/N can be calculated as a function of the laser power. The result is shown in Fig. 3. The calculated S/N ratio exhibits a clear maximum. First the ODMR signal grows linearly with the laser power. Then the size of the ODMR signal becomes constant owing to the saturation of the $^1S_0 \leftarrow {}^1S_1$ transition whereas the light scattering continues to increase. As a result the S/N ratio decreases when the laser power is further increased.

1.6.3 Experimental

Two different experimental arrangements have been successfully utilized for magnetic resonance experiments on single molecular spins [18, 19]. As in the optical single-molecule experiments the essential ingredients are a narrow-band laser, an efficient collection of the emitted fluorescence and a small excitation volume of the sample. To achieve a small sample volume the incoming laser light is either focussed

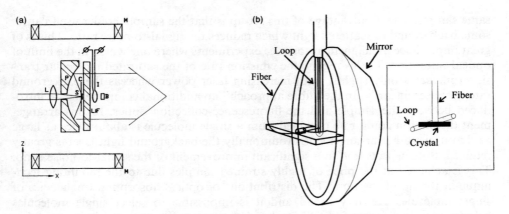

Figure 4. Sketch of the experimental set-up for single-molecule magnetic resonance spectroscopy. In part (a) the sample, S, is mounted between a cover, C, and a LiF substrate, in the joint focus of a parabolic mirror, P, and a lens, L. The microwaves are provided by a loop, I, and the whole sample holder is positioned in the central bore of a superconducting magnet, M. Residual light of the incident laser is blocked by a beam block, B. In part (b) of the figure the sample is mounted at the tip of a single mode fibre, F, centered in the focus of a parabolic mirror, P. Microwaves are provided by a loop, L.

by a small lens close to the sample to a tiny spot ($\approx 5\,\mu$m) or it is guided through an optical fibre which carries the sample at its end. Both experimental configurations are shown in Fig. 4. Part (a) of the figure shows the sample mounted between a LiF substrate and a suprasil cover. The excitation light is focussed onto the sample and the resulting fluorescence is collected with a parabolic mirror. The optical components are adjusted such that their foci coincide on the sample. Moreover the sample holder allows an in situ movement of the sample relative to the excitation light. In part (b) of Fig. 4 it is shown how the sample is mounted on top of the optical fibre which guides the excitation laser light. Again the sample is positioned in the focus of a parabolic mirror which collects the fluorescence. In both arrangements the microwave magnetic field is created by feeding the microwaves into a one-turn loop (2 mm diameter) which is mounted close to the sample. Inhomogeneities of the microwave field are negligible because of the small excitation volume. The microwave-induced change of the fluorescence intensity is monitored in steady state as well as in pulsed microwave experiments by single-photon counting via a digital signal averager or by a correlator to enable photon correlation experiments.

Both constructions feature specific advantages and disadvantages resulting from the sample mounting technique. In the "lens approach" it is possible to mount the crystals with a very low amount of strain. Together with the possibility to move the sample with respect to the focus this allows the selection of a sample volume with an extremely narrow distribution of single-molecule absorption frequencies. As we shall see later this enables the optical selection of single molecules with specific properties. Moreover it is possible to perform single-molecule and ensemble experiments on the same sample. The disadvantage of this set-up is that the sample holder itself causes some background of scattered light which reduces the signal-to-noise ratio. This is of great importance in magnetic resonance experiments where one works in the limit of optical saturation. In this regime the emission rate of the saturated molecular transition increases only slightly with increasing laser power whereas the background rate increases linearly. In the "fibre approach" no additional components are introduced between the sample and the fluorescence-collection optics. In this arrangement the photon count rate obtained from a single molecule is about twice as large as that in the other arrangement. Additionally the background light level is greatly reduced. Both aspects lead to a significant improvement of the signal-to-noise ratio. This is achieved at the price of highly strained samples due to the mounting technique at the tip of the fibre. The distributions of optical absorption frequencies of single molecules are very broad and it is impossible to select single molecules belonging to certain subensembles related to crystal sites or isotopic composition by their optical transition frequency. Moreover the optical transition frequencies of molecules in such samples often show a temporal variation (spectral diffusion) which makes it more difficult to keep the laser in resonance with the molecular transition.

For FDMR experiments it is desirable to be able to apply an external magnetic field. In one of the arrangements the field is provided by a pair of Helmholtz coils outside the cryostat which can be rotated with respect to the sample to obtain the directions of the principal axes of the zero-field tensor. With this approach the exciting laser can be kept focussed on the same sample volume while orienting the field, but the fields are limited to 10 mT. In the other set-up the sample holder is inserted into the bore of a fixed superconducting Helmholtz-type magnet immersed in liquid helium. This allows the application of fields up to 1.3 T, but the direction of the magnetic field with respect to the zero-field tensor cannot be varied.

1.6.4 Mathematical description of the ODMR experiment on single molecules

1.6.4.1 The spin Hamiltonian

The Hamiltonian appropriate for the description of the spin properties of the triplet system is

$$H = \boldsymbol{S} \cdot \mathsf{D} \cdot \boldsymbol{S} + \beta_e g_e \boldsymbol{S} \cdot \boldsymbol{B}_0 + \Sigma_i \boldsymbol{S} \cdot A^{(i)} \boldsymbol{I}_i, \tag{2}$$

with the following meaning:

$\boldsymbol{S}$ –   electron spin operator ($S = 1$)
$\boldsymbol{I}_i$ –   nuclear spin operator of nucleus i
D –   fine structure tensor

β_e –   Bohr magneton of the electron
g_e –   electron g-value
B_0 –   external magnetic field
$A^{(i)}$ –   hyperfine interaction tensor of nucleus i.

The sum in the Hamiltonian runs over all nuclei, intra- and intermolecular, coupled to the electron spin.

The first term in the spin-Hamiltonian, $S \cdot D \cdot S$, causes the lifting of the degeneracy of the three triplet sublevels. The main contribution to this interaction is given by the (magnetic) dipole–dipole interaction of the two unpaired electrons. The resulting energy splitting of the substates is called the zero-field splitting. The operator $S \cdot D \cdot S$ is diagonal in a system of principal axes which correspond to the molecular symmetry axes of pentacene. In this principal axes system (x, y, z) the interaction can be written as

$$H = -XS_x^2 - YS_y^2 - ZS_z^2, \tag{3}$$

with the eigenvalues (energies) $-X$, $-Y$ and $-Z$ of the eigenstates $|X\rangle$, $|Y\rangle$ and $|Z\rangle$, respectively. The eigenvalues fulfill the relation $X + Y + Z = 0$. For the eigenfunctions one obtains the relations $S_u|U\rangle = 0$ ($U = X, Y, Z$) and $S_x|Y\rangle = -S_y|X\rangle = i|Z\rangle$, etc.

The Zeeman term in Eq. (2), $\beta_e g_e S \cdot B_0$, describes the interaction of the electron spin with an externally applied magnetic field. The energies of the spin components depend on the orientation and strength of B_0. The eigenfunctions of $S \cdot D \cdot S + \beta_e g_e S \cdot B_0$ are linear combinations of $|X\rangle$, $|Y\rangle$ and $|Z\rangle$ with coefficients depending on the value and direction of B_0. If B_0 is very large ($g_e \beta_e |B_0| \gg |X|, |Y|, |Z|$) the Zeeman term dominates and the eigenfunctions approximate the eigenfunctions $|S, m_s\rangle$ of S_u. Varying either the direction or the strength of the magnetic field and analyzing the resulting shifts in the triplet transition energies allows the determination of the orientation of the molecular axes system with respect to the magnetic field.

The third part of the spin-Hamiltonian describes the interaction of the electron spin with the surrounding nuclear spins. For the α-protons in pentacene (the protons bound to the carbons in positions α, γ, ε and symmetry equivalent positions, see Fig. 1) the principal axes of the hyperfine tensor coincide with the zero-field tensor axes and the Hamiltonian describing the hyperfine interaction for such a proton can be written $H_{HF} = \delta_i A_{xx}^{(i)} S_x I_x + \delta_i A_{yy}^{(i)} S_y I_y + \delta_i A_{zz}^{(i)} S_z I_z$. Here $A_{xx} = -91$ MHz, $A_{yy} = -29$ MHz, and $A_{zz} = -61$ MHz are empirically determined constants. We have substituted $\delta_i A_{uv}^{(i)}$ for $A_{uv}^{(i)}$ where $A_{uv}^{(i)}$ denotes the uv element of the hyperfine tensor of proton i for a hypothetical spin density of one in the $2p_z$ atomic orbital of the carbon atom to which the proton is bound and δ_i is the actual spin density at the respective carbon nucleus. For the β protons (the protons bound to the carbon in position β and symmetry equivalent positions, see Fig. 1) the in-plane principal axes of the hyperfine tensor are rotated with respect to x and y so to make one of the principal axes parallel to the C–H bond.

The calculation of the lineshapes of the zero-field transitions requires a full diagonalization of the spin hamiltonian. The results of this treatment will be published in a forthcoming paper [20]. Here we approximate the effect of the hyperfine interaction by a second-order perturbation formalism which gives a qualitative insight [21]. The second-order energy shifts are most pronounced for the spin levels $|X\rangle$ and $|Y\rangle$ because they have the smallest energy separation. One obtains for the leading term of the energy shift

$$\Delta E_x = \frac{\frac{1}{4}|\pm \rho_1 A_{zz}^{(1)} \pm \rho_2 A_{zz}^{(2)} \pm \cdots \pm \rho_{14} A_{zz}^{(14)}|^2}{E_x - E_y}$$

$$\Delta E_y = \frac{\frac{1}{4}|\pm \rho_1 A_{zz}^{(1)} \pm \rho_2 A_{zz}^{(2)} \pm \cdots \pm \rho_{14} A_{zz}^{(14)}|^2}{E_y - E_x} \tag{4}$$

$$\Delta E_z \approx 0$$

Each of the 2^{14} nuclear spin basis states, represented by a given selection of the "+" and "−" signs in the sum of Eq. (4) leads to an energy which is slightly shifted from the unperturbed energy. Summing these energies yields a distribution of substate energies within the $|X\rangle$ and $|Y\rangle$ manifold. Due to the quadratic dependence in the numerator in Eq. (4), this leads to asymmetrically shaped magnetic resonance lines for the $|X\rangle$–$|Z\rangle$ and $|Y\rangle$–$|Z\rangle$ transitions.

In the presence of an applied magnetic field the new eigenstates are linear combinations of the zero-field states and the hyperfine interaction may lead to non-vanishing matrix elements in first order. This results in a considerable broadening of the magnetic resonance lines.

1.6.4.2 Kinetic equation of the optical pumping cycle

In this section we discuss the equation of motion of the optical pumping cycle involving the singlet ground state $^1S_0$, the first excited singlet state $^1S_1$ and the three triplet sublevels $|X\rangle$, $|Y\rangle$ and $|Z\rangle$ [18]. The equation allows us to calculate the relative size of the ODMR effect in steady state as well as the auto-correlation function of the fluorescence intensity and the ODMR effect in pulsed experiments. We will see that the equation of motion will describe coherence between the singlet states $^1S_0$ and $^1S_1$, as resulting from the driving laser field, and between the sublevels of the triplet state caused by the resonant microwave field.

The Hamiltonian for the 5-level system subject to a laser field at ω_L resonant with the $^1S_1$–$^1S_0$ transition and a microwave field ω_M resonant with the $|X\rangle$–$|Z\rangle$ transition in the triplet state is written as

$$H = \sum_i \hbar\omega_i |i\rangle\langle i| - \hbar\Omega_R \cos(\omega_L t)(|1\rangle\langle 2| + |2\rangle\langle 1|)$$

$$- \hbar\omega_R \cos(\omega_M t)(|X\rangle\langle Z| + |Z\rangle\langle X|) \tag{5}$$

with the following meaning:

ω_i — energy of state i in units of $\hbar$,

ω_L — laser frequency,

E_0 — amplitude of the laser electric field at frequency ω_L,

i — index which labels the states in the following way:

$\quad i = 1 \quad {}^1S_0$

$\quad i = 2 \quad {}^1S_1$

$\quad i = 3 \quad |X\rangle$

$\quad i = 4 \quad |Y\rangle$

$\quad i = 5 \quad |Z\rangle$,

d_{12} — transition dipole moment between states $i = 1$ and $i = 2$,

Ω_R — Laser Rabi frequency, i.e. $\Omega_R = (E_0 \cdot d_{12})/\hbar$,

ω_M — Microwave frequency

ω_R — Microwave Rabi frequency, i.e. $\omega_R = \gamma B_{1y}$

$\quad \gamma$ — gyromagnetic ratio of the electron,

$\quad B_{1y}$ — the y-component of the microwave magnetic field

The equation of motion in the density matrix description is

$$i\hbar\dot{\rho} = [H, \rho]. \tag{6}$$

On rewriting the equation in a reference frame rotating at the resonance frequency of the optical and microwave transition and neglecting all coherence effects between singlet and triplet states we find an equation of motion of the form

$$\frac{dP}{dt} = R \cdot P + S \tag{7}$$

with

$$
S = \begin{bmatrix} 0 \\ 0 \\ \dfrac{\Omega_R}{2} \\ 0 \\ 0 \\ 0 \\ 0 \\ 0 \end{bmatrix}
\qquad
P = \begin{bmatrix} \sigma_{22} \\ u \\ v \\ \sigma_{xx} \\ \sigma_{yy} \\ \sigma_{zz} \\ s \\ t \end{bmatrix}
\tag{8}
$$

Here σ_{22}, σ_{xx}, σ_{yy}, σ_{zz} denote the populations of ${}^1S_1$, $|X\rangle$, $|Y\rangle$ and $|Z\rangle$. The terms u, v denote the coherence between ${}^1S_1$ and ${}^1S_0$ and s and t the coherence between the triplet sublevels $|X\rangle$ and $|Z\rangle$ caused by the presence of the resonant microwave field. The relaxation matrix R is defined by

$$
R = \begin{bmatrix}
-K_2 & \Omega_R & 0 & 0 & 0 & 0 & 0 \\
 & -\Gamma_{12} & \delta_L & 0 & 0 & 0 & 0 & 0 \\
-\Omega_R & \delta_L & -\Gamma_{12} & -\dfrac{\Omega_R}{2} & -\dfrac{\Omega_R}{2} & -\dfrac{\Omega_R}{2} & 0 & 0 \\
k_{x2} & 0 & 0 & -k_{1x} & 0 & 0 & 0 & -\omega_R \\
k_{y2} & 0 & 0 & 0 & -k_{1y} & 0 & 0 & 0 \\
k_{z2} & 0 & 0 & 0 & 0 & -k_{1z} & 0 & \omega_R \\
0 & 0 & 0 & 0 & 0 & 0 & -\Gamma_{xz} & \delta_M \\
0 & 0 & 0 & \dfrac{\omega_R}{2} & 0 & -\dfrac{\omega_R}{2} & -\delta_M & -\Gamma_{xz}
\end{bmatrix} \tag{9}
$$

Here δ_L denotes the difference between the frequency of the optical field and the resonance frequency of the $^1S_1 \leftarrow {}^1S_0$ transition and δ_M the difference between the frequency of the microwave field and the resonance frequency of the $|X\rangle$–$|Z\rangle$ transition. Γ_{12} and Γ_{xz} are the dephasing rates of the optical and the microwave transitions, k_2 is the total decay rate of $^1S_1$. k_{1x}, k_{1y}, k_{1z} are the depopulating rates of the sublevels of $^3T_1$ and k_{x2}, k_{y2}, k_{z2} their populating rates.

The steady-state populations for given optical and microwave powers can be deduced from the solution of the linear system

$$
\frac{dP}{dt} = 0 \Rightarrow R \cdot P_\infty = -S \tag{10}
$$

The time dependence of the populations after sudden preparation in a given state described by P_0 follows from

$$
P(t) = e^{Rt} \cdot (P_0 - P_\infty) + P_\infty \tag{11}
$$

The solutions of Eqs. (10) and (11) may be found by numerical diagonalization of Eq. (9).

1.6.4.3 The correlation function of the fluorescence intensity and the influence of microwave irradiation

The photon autocorrelation function of single molecules has been shown to be a powerful tool for investigating the kinetics of the optical pumping cycle [15]. This fluorescence autocorrelation function

$$
g^{(2)}(\tau) = \frac{\langle I(t)I(t+\tau)\rangle}{\langle I(t)\rangle^2}, \tag{12}
$$

is discussed in detail in other chapters of this book. Here $I(t)$ denotes the fluorescence intensity at time t. This process is supposed to be stationary and the average is over a time much longer than the characteristic times of the intensity fluctuations. In this model the probability of detecting a pair of photons in intervals $(t, t + dt)$ and $(t + \tau, t + \tau + dt)$ is proportional to the probability of occupation of the excited singlet state at time t and the conditional probability that the molecule will be in this state at $t + \tau$. This conditional probability is just the matrix element $\sigma_{22}^0(\tau)$ corresponding to the solution of Eq. (11) with $\boldsymbol{P}_0 = 0$. The correlation function within the three-level model, neglecting coherence in the $^1S_1 \leftarrow {}^1S_0$ transitions and in the absence of a resonant microwave field has been calculated analytically [15]. When neglecting spectral diffusion processes $g^{(2)}(\tau)$ can be written as

$$g^{(2)}(\tau) = 1 + ce^{-\lambda t}, \tag{13}$$

with

$$\lambda = k_{13} + \frac{k_{32}\Omega_R^2}{2(k_2\Gamma_{12} + \Omega_R^2)}, \tag{14}$$

and

$$C = \frac{\lambda - k_{13}}{k_{13}}, \tag{15}$$

In these equations k_{32} and k_{13} are respectively the effective intersystem crossing rates from $^1S_1$ to $^3T_1$ (viewed as a single level) and from $^3T_1$ to $^1S_0$. At high power ($\Omega_R \to \infty$) $\lambda \to k_{13} + k_{32}/2 \approx k_{32}/2$ for pentacene and $C = k_{32}/2k_{13}$, while at low power ($\Omega_R \to 0$), $\lambda \to k_{13}$ and $C \to 0$. Hence the intersystem crossing rates may be found from experimental correlation functions. The contributions of these parameters to $g^{(2)}(\tau)$ can be modified by the coupling of the triplet sublevels with a resonant microwave field. In this case one has to use the description of $g^{(2)}(\tau)$ for the coupled 5-level system [18].

1.6.5 Steady-state magnetic resonance

1.6.5.1 Electron-paramagnetic resonance

In the following the magnetic resonance of single pentacene ($C_{22}H_{14}$) molecules doped into a p-terphenyl ($C_{18}H_{14}$) crystal will be discussed for steady-state microwave excitation conditions. If the isotopic constitution of a molecule is not stated explicitly it corresponds to exclusively ^{12}C and ^1H nuclei. Owing to the four distinct lattice sites of p-terphenyl at low temperature four spectral origins, termed O_1 to O_4, show up in the absorption spectrum. Single molecules with their absorption frequencies close to the O_1 and O_2 origins are optically isolated by the methods described elsewhere in this book.

Even in zero-magnetic field the triplet state of pentacene is split into the three zero-field eigenstates $|X\rangle$, $|Y\rangle$ and $|Z\rangle$. To observe the transitions between these states of

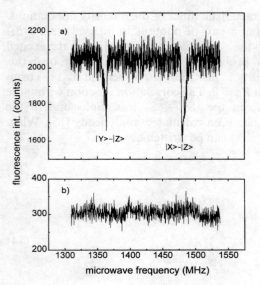

Figure 5. (a) ODMR spectrum of a single molecule. The sampling time for this spectrum is about 600 s. The optical excitation intensity has been chosen to achieve a maxium S/N ratio for the ODMR signal. (b) Fluorescence intensity when the laser frequency is detuned from the fluorescence–excitation line of a single molecule.

a single molecule the laser is tuned to the peak of the single molecule fluorescence excitation line and the power is adjusted to saturate the optical $^1S_1 \leftarrow {}^1S_0$ transition. Subsequently the fluorescence intensity is recorded as a function of the microwave frequency. In Fig. 5(a) the $(|Y\rangle - |Z\rangle)$ and the $(|X\rangle - |Z\rangle)$ magnetic resonance transitions are observed as a decrease (up to 25%) of the fluorescence. This is caused by the increased population probability of the long-lived $|Z\rangle$ level, which lengthens the dark intervals between the photon bunches and consequently reduces the time-averaged fluorescence intensity. The third transition $(|X\rangle - |Y\rangle)$ is much weaker due to unfavourable populating and depopulating kinetics. In Fig. 5(b) it is verified that the signal vanishes if the laser is out of resonance with the optical transition of the molecule.

Fig. 6 compares the lineshapes of the $(|Y\rangle - |Z\rangle)$ transition for a single molecule and an ensemble of about 10^4 molecules. The ensemble spectrum was obtained with the laser in resonance with the fluorescence excitation line of the O_1 ensemble. The transition shows an asymmetric lineshape with a steep decrease towards higher microwave frequencies for both the single molecule and the ensemble case. The lineshape results from the hyperfine interaction of the triplet electron spin with the pentacene proton spins ($I = 1/2$). Each proton can exist in one of its two nuclear spin states which yields 2^{14} nuclear spin configurations. The hyperfine interaction of each of these nuclear configurations causes a slight shift of the resonance. As pointed out in Section 4.1. In zero-field the hyperfine interaction is a second-order effect which leads to the observed opposite asymmetric lineshapes for the $(|Y\rangle - |Z\rangle)$ and the $(|X\rangle - |Z\rangle)$ transitions (see Fig. 5).

For a single molecule one would expect that it "sees" only one nuclear spin configuration and that a very narrow magnetic resonance line would be observable. Apparently, it experiences all of these configurations during the many optical

Figure 6. Comparison of the lineshapes of the $|Y\rangle$–$|Z\rangle$ magnetic resonance transition for a single molecule (top) and for an ensemble of about 10^4 molecules (bottom). In contrast to the previous figure the vertical scale corresponds to a decrease of fluorescence. The ensemble spectrum was recorded with the laser tuned to the top of the inhomogeneously broadened O_1 absorption line whereas the single molecule was selected in the red wing of this line.

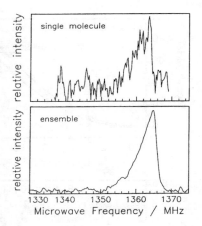

pumping cycles which are needed to accumulate a sufficient signal-to-noise ratio. This is due to the dipolar coupling among the proton spins which leads to a spin diffusion within the proton reservoir of the guest and the host. When the triplet magnetic moment is created, the 14 proton spins on the pentacene suddenly "feel" the (second-order) hyperfine fields which shift their resonance frequency away from the dipolar spectrum of the protons in the bulk of the crystal. Consequently, during the triplet lifetime, this configuration is frozen and the resonance frequency can only vary in a small interval $\Delta\nu$ determined by the flip–flop motions of the protons in the bulk. This interval can be estimated from the electron spin–spin relaxation time T_2 and amounts to $\Delta\nu = 1/\pi T_2 \approx 150\,\text{kHz}$ [14, 22]. On return to the ground state, the hyperfine fields disappear and the pentacene protons are free to participate in the nuclear flip–flop motion. When the molecule is excited again into the triplet state, a new magnetic configuration is frozen, which corresponds to a different position in the zero-field resonance line. An estimate of the related time-scales yields that the average time between two excitations into the triplet state is about 20 µs and that the mean residence time of the molecule in the triplet state is about 50 µs. For the inverse of the flip–flop rate one can estimate a value of about 30 µs which means that each time the molecule reappears in the triplet state it experiences a different nuclear configuration. Thus the resemblance of the ensemble and the single molecule spectrum of Fig. 6 reflects the validity of the ergodic theorem for this case. The time average of an observable for a single molecule is equivalent to the ensemble average of that observable.

As yet the ODMR experiments have been described without an external B_0 field. Fig. 7 shows how the application of a B_0 field affects the line position as well as the width of the two ODMR transitions for pentacene in p-terphenyl. Both transitions are shifted to higher frequencies and the lines are broadened considerably. The shift of the ODMR line depends on the magnitude of B_0 (Zeeman effect) as well as on its orientation with respect to the principal axes system of the fine-structure tensor. As can be seen in Fig. 7(a) the experimentally observed shifts (triangles) agree well with the calculated curves. The line broadening (30 MHz in Fig. 7(b) is a consequence of

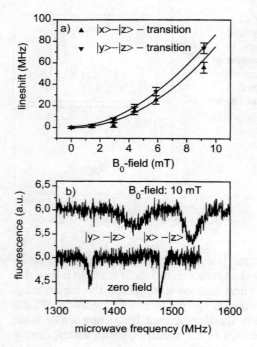

Figure 7. (a) Shift of the ODMR transition frequencies as a function of the field strength of the external magnetic field. The triangles represent measured values and the solid curve is a result of a simulation according to Eq. (2). (b) Comparison of the ODMR lineshape with and without magnetic field. The field strength is 10 mT. The linewidth of the $|X\rangle$–$|Z\rangle$ and $|Y\rangle$–$|Z\rangle$ transition in this case is roughly 30 MHz.

the fact that the hyperfine interaction between the electron spin and the protons changes from a second-order to a first-order effect. It is interesting to note that already for fields of 10 mT and for particular orientations of B_0 with respect to the principal axes of the fine-structure tensor the expectation value of S_z reaches values up to 0.8 due to the fact that the electron spin is nearly quantised in the direction of the external magnetic field. As far as the linewidth is concerned one almost reaches the high-field limit.

By performing a systematic variation of the orientation of the external magnetic field with respect to the sample one is able to determine the orientation of the principal axes system of the fine-structure tensor of an individual molecule with respect to the laboratory reference frame. Since for symmetric molecules such as pentacene this principal axes system coincides with the molecular symmetry axes, one thus can determine the orientation of individual molecules even in macroscopically random samples.

In Fig. 8(a) we show the experimental result of an orientational variation of B_0 in the x', y' plane of the laboratory frame (see Fig. 8(b)) for 2 different molecules (dots and triangles) together with simulated curves. This experiment was performed in the setup where the sample is mounted at the tip of the fibre. Owing to the induced crystal strain the optical transitions of sites O_1 and O_2 then overlap and it is impossible to decide to which site a molecule belongs. It is seen that both molecules show a rather different behaviour of the line position as a function of the orientation of B_0. Whereas M_1 undergoes a marked change of the ODMR transition frequency, M_2 hardly shows any variation. This is a signature for a different orientation of

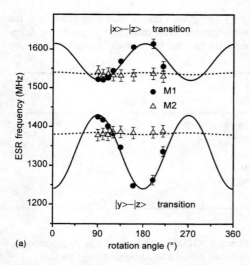

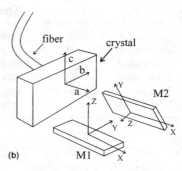

(a)   (b)

Figure 8. (a) Dependence of the ODMR transition frequencies on the angle of rotation of the magnetic field in the $x'y'$ plane of the laboratory axis system, for two different molecules M_1 (filled circles) and M_2 (triangles). Symbols represent measured values whereas the solid and dotted lines are the result of a fit using Eq. (2). (b) Schematic sketch of the sample at the tip of the fibre in the x', y', z' laboratory frame. The two molecules M_1 and M_2 are depicted schematically and their molecular axes systems x, y, z are indicated. Molecule M_1 has been placed arbitrarily to lie in the $x'y'$ laboratory plane whereas the relative orientation of molecule M_2 with respect to M_1 results from the analysis of the data presented in part (a) of the figure.

both molecules. A closer analysis reveals, that the molecules occupy either the sites (P_1/P_2) or (P_3/P_4) of the p-terphenyl crystal [23]. Both molecules are rotated by about 60° about the long molecular x-axis with respect to each other and form an angle of 20° with regard to the normal of the crystal plane. All molecules investigated in this crystal belong to either one of this class. The experimental accuracy of $\pm 4°$ [24] precludes the observation of small angular variations of molecules belonging to another class.

Table 1 compares the ODMR transition frequencies for four different molecules with the data for the O_1 and O_2 ensembles. It appears that for three of the four molecules (M_1, M_2, M_3) the ODMR transition frequencies and the mutual orientation are consistent with the ensemble data for O_1 and O_2. For the fourth molecule this is not the case.

1.6.5.2 Hyperfine interaction of a single molecular spin with individual nuclear spins

To study the hyperfine interaction of a single molecular spin with individual nuclear spins one has to create a situation in which the hyperfine interaction with one or two nuclear spins dominates. This can be achieved by using isotopically labeled molecules such as pentacene in which all the protons have been replaced by deuterium except for the central positions. The magnetic moment of deuterium ($I = 1$) is more than a factor of 6 smaller than that of a proton ($I = 1/2$). Since the electron spin

Table 1. Orientation of four pentacene molecules ($\pm 4°$) together with the transition energies of the two zero-field transitions. Θ denotes the angle between the individual molecular y axes of the different molecules. For comparison the corresponding data for the O_1 and O_2 ensembles are also included.

	$\Theta/°$	$\|X\rangle - \|Z\rangle\|$ MHz	$\|Y\rangle - \|Z\rangle$ MHz
ensemble O_1	0°	1480.3	1363.2
ensemble O_2	61°	1479.3	1360.0
Molecule 1	0°	1480.5	1363.5
Molecule 2	61°	1478.5	1360.5
Molecule 3	0°	1481.2	1362.0
Molecule 4	0°	1479.6	1361.0

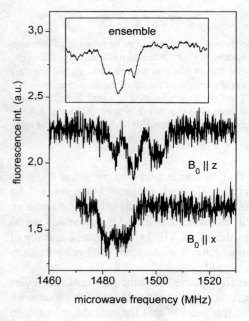

Figure 9. ODMR lineshape of a pentacene-$h_2 d_{12}$ molecule where all hydrogens except the two bound to carbon in the γ position (see Fig. 1) are replaced by deuterons. The upper trace shows the $\|X\rangle - \|Y\rangle$ transition when B_0 is applied along the molecular z axis whereas for the lower trace the magnetic field is applied along the molecular x direction. For comparison the inset shows an ensemble spectrum.

density is larger at the central carbon positions the hyperfine interaction with the two central protons dominates. In the presence of a magnetic field a three-line spectrum is expected and indeed observed experimentally, Fig. 9. The field strength in this example is approximately 3 mT and the orientation of the field is roughly parallel to the molecular Y-axis. The splitting amounts to about 8 MHz and the intensity ratio of the three lines is, as expected, $1:2:1$. The center line is the transition between those levels when both nuclear spins are antiparallel whereas the outer lines belong to those transitions where both spins are parallel but in one case aligned parallel to the "effective" field and in the other case antiparallel. The effective field corresponds

Figure 10. Fluorescence excitation spectrum of the O_1 spectral site of pentacene in p-terphenyl. The satellites 1–5 result from pentacene molecules which contain a single ^{13}C nucleus (natural abundance).

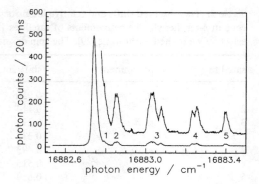

to the vector sum of the external field B_0 and the hyperfine field caused by the electron spin. Since the Zeeman energy of the electron spin in the present case is smaller than or at best comparable to the fine-structure splitting the two field directions do generally not coincide. The experimentally determined coupling constant is in agreement with the reported hyperfine interaction for this specific molecular position.

The spectral spin diffusion rate caused by nuclear spin flip processes in the bulk nuclear spin system in the sample pentacene-$h_2 d_{12}$ in p-terphenyl-d_{14} is slower than in the case of a normal, protonated mixed crystal [25]. Nevertheless all three possible configurations of the proton nuclei appear in the spectrum implying that the effective spin diffusion time is still much shorter than the time of 10 minutes needed to record the spectrum.

Hyperfine interaction of a single triplet spin with a single nuclear spin can be observed without resorting to specifically labeled compounds, by optically selecting pentacene molecules which contain rare isotopes in natural abundance. This is illustrated in Fig. 10 which shows the optical excitation spectrum of a sample of pentacene in p-terphenyl, carefully mounted between a LiF substrate and a suprasil cover to prevent strain in the crystal. The high quality of the crystal is illustrated by the extremely narrow, inhomogeneously broadened O_1 ensemble line. The linewidth of 750 MHz (FWHM) is only a factor of 100 larger than the homogeneous linewidth, which allows the observation of a set of weak satellite features, numbered 1 to 5, in the high-energy wing. These satellites are caused by pentacene molecules which contain a single ^{13}C nucleus [26, 27]. The substitution with a heavier nucleus causes a decrease of the vibrational energies and consequently of zero-point energies in the ground state and to a lesser extent in the excited state which results in a slight blue shift of the optical absorption frequency for such molecules. Given the small linewidth it is not surprising that such satellites are observed because the probability of finding a pentacene molecule containing a ^{13}C atom (natural abundance 1.108%) is no less than 19.2%. For a pentacene molecule of D_{2h} symmetry a ^{13}C isotope can occupy six inequivalent positions within the molecule and one expects to observe six satellites with an intensity ratio of $4:4:4:4:4:2$. As can be seen from Table 2 the intensity distribution among the satellites agrees with this picture. The small splitting of satellite 4 is probably caused by a slight deviation from D_{2h} symmetry which is consistent with the crystal site having only inversion symmetry. From magnetic res-

Table 2. The observed shifts and relative intensities of ^{13}C satellites for the O_1 spectral site of pentacene in *p*-terphenyl. The assignments of satellites 1 to 3 are based on a comparison with studies on selectively enriched anthracene [29]. The spin densities are taken from Ref. 30.

satellite	assignment	shift/cm^{-1}	rel. intensity/a.u.	spin density
1	β	0.056	4	0.025
2	ζ	0.114	4	−0.015
3	α, δ	0.292	8	0.045 (α)
		0.337		−0.021 (δ)
4	ε	0.493	4	0.128
		0.515		
5	γ	0.659	2	0.188

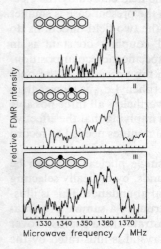

Figure 11. Comparison of the lineshapes of the $|Y\rangle$–$|Z\rangle$ magnetic resonance transitions for three different molecules. The insets show the position of the ^{13}C substitution as obtained from the analysis of the hyperfine interaction.

onance experiments it could be concluded that satellites 4 and 5 result from pentacene molecules containing a ^{13}C nucleus in position ε or γ, respectively (see Fig. 1) [27, 28]. In table 2 the relevant data of the satellites are summarized. Assignments other than for satellites 4 and 5 are based on similar conclusions from a study of selectively ^{13}C enriched anthracene [29].

The fluorescence excitation spectrum, Fig. 10, suggests that single molecules without ^{13}C are most likely to be encountered to the red of the main line, whereas single molecules containing a ^{13}C nucleus should be abundant with high probability between the satellite lines. In Fig. 11 the FDMR spectra in zero-field of the ($|Y\rangle$–$|Z\rangle$) transition for three different molecules are shown. The location of the optical absorption for the three molecules was: to the red of the main line for molecule I, in between satellites 4 and 5 for molecule II and to the blue of satellite 5 for molecule III. The insets in each figure depict the position of the ^{13}C substitution of each molecule. Whereas molecule I contains ^{12}C nuclei exclusively, molecule II contains a ^{13}C

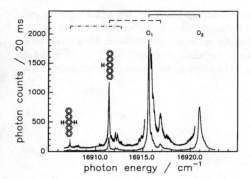

Figure 12. Fluorescence–excitation spectrum of pentacene-d_{14} in p-terphenyl-d_{14}. The strong lines, labelled O_1 and O_2, connected by the brackets, correspond to pentacene-d_{14} molecules in the O_1 and O_2 spectra sites of p-terphenyl. The O_2 line appears weaker in the spectrum because the polarization of the incident laser was adjusted to give a maximum signal for the O_1 line. The two lines connected by the dashed bracket correspond to the O_1 and O_2 ensemble lines of pentacene molecules which contain a single proton bound to the carbon in the γ position. Similarly the two lines connected by the dashed–dotted bracket result from pentacene molecules containing two protons each of them bound to a carbon in the γ position of pentacene.

nucleus in the ε position and molecule III contains one in the γ position (see Fig. 1). The ($|Y\rangle$–$|Z\rangle$) transition lines for molecules II and III show a considerable broadening with respect to the line obtained for molecule I. This is caused by the additional hyperfine interaction with the ^{13}C nuclear spin ($I = 1/2$). The one-center contribution to the ^{13}C hyperfine interaction amounts to $A_{xx}^{(C)} = 34$ MHz, $A_{yy}^{(C)} = 34$ MHz and $A_{zz}^{(C)} = 307$ MHz for a spin-density of one in a $2p_z$ atomic orbital. Taking into account the known spin-density distribution of the triplet state of pentacene [30] and following a second-order perturbation treatment as outlined in Section 1.6.4.1 the positions of the ^{13}C substitutions for molecules II and III could be assigned by comparing the calculated broadenings with the experimentally observed ones.

Optical selection of isotopically substituted single molecules is also possible in samples that are isotopically enriched. In Fig. 12 the fluorescence-excitation spectrum of pentacene-d_{14} in p-terphenyl-d_{14} is shown. The two strong, inhomogeneously broadened lines labeled O_1 and O_2 correspond to ensembles of pentacene-d_{14} occupying the O_1 and O_2 spectral sites of the p-terphenyl-d_{14} crystal. The degree of deuteration of the pentacene was 98% which means that pentacene molecules which contain one or two protons are still abundant in the sample. At position $16\,911.41$ cm^{-1} a relatively weak fluorescence–excitation line is found which is identified as the O_1 ensemble line of molecules which contain one proton bound to the carbon in the γ position. At position $16\,907.26$ cm^{-1} an even weaker line is found which is assigned to pentacene molecules in the O_1 spectral site which contain two protons each bound to a γ carbon. Similarily their O_2 counterparts could be found in the spectrum. The assignment of the positions of the proton substitutions is based on the analysis of FDMR experiments performed on these (small) ensembles and ana-

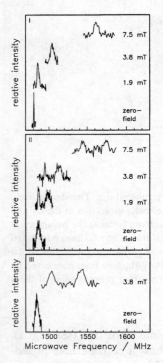

Figure 13. Fluorescence-detected magnetic resonance signal of the $|X\rangle - |Z\rangle$ transition for three different pentacene-d_{14} molecules. Molecules I and II are O_1 molecules and molecule III is an O_2 molecule. Whereas molecule I consists completely of ^{13}C nuclei molecules II and III contain a ^{13}C nucleus in the γ position of pentacene.

lysing their hyperfine interaction. Taking a closer look at the spectrum reveals a variety of lines caused by pentacene molecules containing one or two protons in various positions of the molecule. Concerning ^{13}C this sample corresponds to natural abundance which means that additionally combinations of ^{13}C and one or two proton substituted molecules give rise to weak ensemble lines.

In Fig. 13 FDMR spectra of the $|X\rangle - |Z\rangle$ transition are shown for three different pentacene-d_{14} molecules containing no or one ^{13}C nucleus, both in zero-field and in the presence of a weak magnetic field. The molecules are selected optically in a similar way as in the protonated samples discussed above. Molecules I and II are O_1 molecules while molecule III is an O_2 molecule excited to the blue of the O_2 ensemble line. Molecule I consists completely of ^{12}C nuclei whereas molecules II and III contain a single ^{13}C nucleus in the center position. For molecule I one observes in zero-field a very narrow magnetic resonance line due to the small deuterium hyperfine interaction. Upon application of an external magnetic field up to 7.5 mT a shift and a broadening of the transition is observed. The shift results from the Zeeman interaction of the triplet electron spin with the external magnetic field while the broadening results from the hyperfine interaction with the 14 deuterons which turns from a second-order to a first-order effect when a magnetic field is applied. The magnetic resonance transition of molecule II shows, besides a frequency shift of the transition similar to that observed for molecule I, a splitting in the presence of a magnetic field. This splitting is caused by the hyperfine interaction of the triplet spin with the single

Table 3. Comparison of the calculated and observed hyperfine splitting for various molecules. The labels ε and γ refer to the position of the ^{13}C substitution.

Molecule (type)	^{13}C position	Magnetic Field (mT)	Observed Splitting (MHz)	Calculated Splitting (MHz)
II (O_1)	γ	1.9	12.3	11.2
		3.8	18.0	20.7
		7.5	30.0	31.9
III (O_2)	γ	3.8	39.0	43.8
IV (O_1)	γ	3.8	19.0	20.7
V (O_1)	γ	3.8	19.0	20.7
VI (O_1)	ε	1.9	7.6	8.0
		3.4	11.0	13.4
VII (O_2)	γ	3.8	39.0	43.8
VIII (O_2)	γ	3.4	42.0	41.8

^{13}C nuclear spin. For molecule III both the observed shift and the splitting of the magnetic resonance transition are different from those obtained for the O_1 molecules owing to the different orientation of O_1 and O_2 molecules with respect to the magnetic field. Utilizing a spin-Hamiltonian as given in Eq. (2), where for the hyperfine interaction only the ^{13}C nuclear spin is taken into account (thus neglecting the deuterium hyperfine and quadupole interaction) and using the known orientation of the magnetic field with respect to the molecule under study, the splittings of the magnetic resonance lines have been calculated [31]. The result is summarized for various molecules together with the experimental data in Table 3. The observed splitting demonstrates that the flip–flop time of the single ^{13}C spin is short compared to the observation time of 20 minutes. In this aspect the situation is similar to the case of pentacene-h_2d_{12} where three hyperfine lines of the two ^{1}H nuclear spins were observed.

From the experiments it has become evident that the extremely high sensitivity of single molecule spectroscopy enables the study of isotopically substituted molecules – either synthesized or in natural abundance – in the form of single molecules or as ensembles. Magnetic resonance spectroscopy provides a unique tool to identify these isotopic modifications owing to the specificity of the hyperfine interaction.

1.6.5.3 Determination of the intersystem crossing rates via the autocorrelation function of the fluorescence intensity

Fig. 14 compares the decay of the photon correlation function of a single pentacene molecule with and without a microwave field resonant with the $|X\rangle$–$|Z\rangle$ transition. It is obvious that the contrast, C, is strongly influenced by the presence of the microwave field. It is defined in Eq. (15) as the ratio of population versus depopulation rates of the triplet state. The increase of the contrast in the presence of the resonant microwaves results from the accompanied reduction of the triplet depopulation

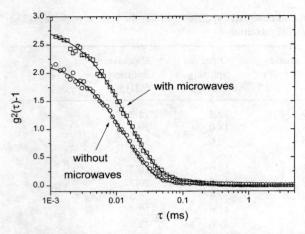

Figure 14. The autocorrelation function of the fluorescence intensity $g^{(2)}(\tau)$ with and without the irradiation of microwaves in resonance with the $|X\rangle-|Z\rangle$ transition. The diamonds and the crosses represent measured values and the solid curve is the result of a simulation according to Eq. (13).

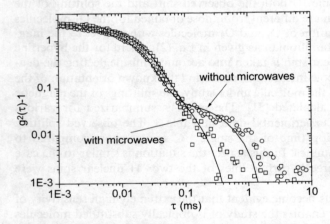

Figure 15. Double-logarithmic plot of $g^{(2)}(\tau)$ for irradiation at various microwave powers (resonant with the $|X\rangle-|Z\rangle$ transition). Points are measured values and the solid line is a simulation according to Eq. (10).

rate. The decay of $g^{(2)}(\tau)$, however, remains nearly unaltered by the presence of the microwaves because in the case of pentacene it is mainly determined by the population rate of the triplet state, Eq. (14). The decay of $g^{(2)}(\tau)$ is not mono-exponential, a feature characteristic for a 5-level system, as is more clearly seen in a double-logarithmic plot, Fig. 15. The full lines correspond to a numerical fit of the results to Eq. (14). The fit parameters give the complete set of populating and depopulating rates of one distinct pentacene molecule and are summarized in Table 4. The second exponential in the decay of $g^{(2)}(\tau)$ results from the population of the

Table 4. ISC parameter of a single pentacene molecule as determined from the analysis of $g^{(2)}(\tau)$ and pulsed microwave experiments.

Symbols	Meaning	Value/MHz
k_2	fluorescence decay	43 [46]
Γ_{12}	optical dephasing	25 [47]
Γ_{XZ}	microwave dephasing	1.2
Ω_R	optical Rabi frequency	92
ω_R	microwave Rabi frequency	15
k_{X2}	population rate of $|X\rangle$	0.066
k_{Y2}	population rate of $|Y\rangle$	0.029
k_{Z2}	population rate of $|Z\rangle$	0.00028
k_{1X}	depopulation rate of $|X\rangle$	0.02
k_{1Y}	depopulation rate of $|Y\rangle$	0.02
k_{1Z}	depopulation rate of $|Z\rangle$	0.0012

long lived $|Z\rangle$ level. For low microwave power the decay of this component reflects directly its lifetime whereas the amplitude indicates its small population probability. As can be seen both rates are influenced by the microwaves [18].

1.6.6 Spin coherence experiments

1.6.6.1 Transient nutation

The simplest coherence experiment in the case of single spin EPR, is a (transient) nutation experiment [22]. In this experiment the spin is forced to precess perpendicular to the direction of the microwave magnetic field component in the rotating frame. This is equivalent to a periodic modulation of the population probability of the two $^3T_1$ sublevels in resonance with the microwaves. This periodic modulation leads to a change in the lifetime of the $^3T_1$ state and thus to a corresponding modulation of the average fluorescence intensity.

Fig. 16 shows the experimental result of a nutation experiment on a single triplet electron spin for two different microwave powers. In this experiment the average fluorescence intensity is recorded as a function of the duration of the pulse length. The figure clearly shows a periodic modulation of the fluorescence intensity as it is also found in a conventional pulsed ODMR experiment on a spin ensemble. The fluorescence intensity starts at a high level, corresponding to a large population probability of the short-lived $|X\rangle$ sublevel. When increasing the pulse length (up to 100 ns) it gradually approaches a low fluorescence intensity equivalent to a maximal population probability of the long-lived $|Z\rangle$ sublevel. For a still longer pulse length the situation reverses and the spin rotates back to its starting position before a new nutation period begins. Naively one might expect only two distinct fluorescence intensities to appear, resulting either from the population of the $|X\rangle$ or $|Z\rangle$ sublevel. For sensitivity reasons the molecule has been pumped several 10^6 times into the triplet state to record the experimental trace in Fig. 16. As a result one measures

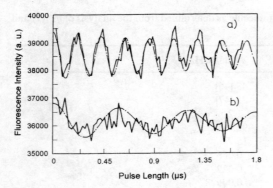

Figure 16. Transient nutation patterns for two different microwave powers. For trace (b) the microwave power was attenuated by 20 dB with respect to the microwave power used to record trace (a). The dotted lines represent a simulation according to Eq. (16).

the time averaged population probabilities of the $|X\rangle$ and $|Z\rangle$ level. According to the ergodic theorem the single spin thus mimics an ensemble behavior.

Due to experimental limitations only nutation traces up to a duration of 2 μs could be recorded. Within this time interval the modulation amplitude of the nutation is hardly damped. This is in sharp contrast to EPR ensemble experiments, where a damping of the nutation usually is clearly visible within the first μs. The damping is caused by an inhomogeneity of the driving microwave field across the sample. This of course is absent in the single-molecule experiment, since the field is homogeneous over a molecular dimension. The nutation experiment can be described by [32, 33],

$$r_3(t) = r_3(0)\frac{1}{\omega_{\text{eff}}^2}(\delta_{\text{M}}^2 + \omega_{\text{R}}^2 \cos \omega_{\text{eff}} t) \qquad (16)$$

r_3 is the population difference between the two triplet sublevels involved and ω_{eff} is given by $\omega_{\text{eff}} = (\omega_{\text{R}}^2 + \delta_{\text{M}}^2)^{1/2}$. For $\omega_{\text{R}} \gg \delta_{\text{M}}$, $r_3(t)$ describes an undamped nutation whereas for smaller ω_{R} the nutation is damped. This damping, determined by the ratio of ω_{R} and the inhomogeneous linewidth, should also be observable in a single-spin experiment and is considered in the simulation in Fig. 16. However, due to the short time range no experimental evidence for this effect can be found.

1.6.6.2 Hahn echo

As has already been discussed in a previous section the single-molecule ODMR line for pentacene is inhomogenously broadened. The Hahn echo (HE) is a pulse sequence known in magnetic resonance to suppress the effect of static inhomogeneities. During the lifetime of the $^3T_1$ state a given frequency position of the ODMR transition within the inhomogenously distributed density of states is fixed. From the point of view of pulsed ODMR experiments which are only effective when the molecule is in the $^3T_1$ state this inhomogenous distribution is thus static. Following the argument of ergodicity it can be anticipated that the single molecule coherence observed in the proceeding paragraph should be refocussable by a two-pulse echo sequence when averaging in time. Optical detection of spin coherence in triplet states is possible by the addition of a third (probe) pulse to the conventional two-pulse HE sequence [34], see inset Fig. 17, which projects the transverse magnetization into an

Figure 17. The Hahn Echo of a single molecular spin. The figure shows the normalized change in fluorescence intensity as a function of τ_2, the separation between the second and the third pulse as shown in the inset of the figure. The circles represent measured values whereas the solid line is the result of a simulation using Eq. (11).

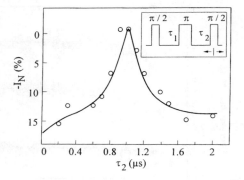

optically detectable population difference. An echo can be detected by changing the time interval τ_2 between the second and the third pulse while observing the change in fluorescence intensity. The experimental evidence of a single-spin echo is shown in Fig. 17. The figure shows the normalized change in fluorescence intensity as a function of τ_2. It reveals the expected behavior of the fluorescence intensity when a spin echo is measured. The solid curve represents a simulation with a simple density-matrix treatment already described in ref. [18]. The width and the height of the echo is well reproduced by this treatment, an additional hint, that the observed structure indeed is the result of a HE on a single triplet electron spin. The asymmetry at short times τ_2 is due to an overlap of the HE with the free-induction decay, which is also included in the simulation.

The existence of a single-spin echo is an interesting phenomenon. In addition it allows to determine the homogeneous linewidth of the EPR transition by measuring the decay time T_2 of the HE amplitude. Comparison of the homogeneous optical linewidth of different chromophores -especially in disordered solids- have shown an appreciable dispersion of this value, which can be used to characterise the host matrix [35, 36]. Measurements of the homogenous linewidth of the EPR transition thus might be an additional means to derive information about the dephasing mechanisms of the electron spin and the structure of the host matrix.

The "dephasing" of a single electron spin is measured via the decrease of the amplitude of the fluorescence intensity after the probe pulse (see Fig. 18). The figure shows the temporal evolution of the fluorescence intensity after the third pulse which returns back to equilibrium with the time constant of the long lived $|Z\rangle$ level after the increase in fluorescence intensity due to the HE pulse sequence. For short delay times τ the dephasing is negligible and the effect of the three pulses is to rotate $r_3(t)$ by 2π. Thus the fluorescence intensity after the third pulse remains unaltered (see trace for $2\tau = 0.2\,\mu s$, Fig. 18). However, for large τ dephasing mechanisms become active and a change in the fluorescence intensity after the HE pulse sequence shows up (see trace for $2\tau = 10\,\mu s$, Fig. 18). The treatment described in [18] is able to reproduce this effect as it is shown by the solid lines in Fig. 18.

A typical decay of the HE of a single molecule is given in Fig. 19. This figure shows the HE intensity as a function of τ_1 ($\tau_1 = \tau_2$), where τ_1 is the waiting time between the first two pulses. The dotted curve represents a fit with a mono-

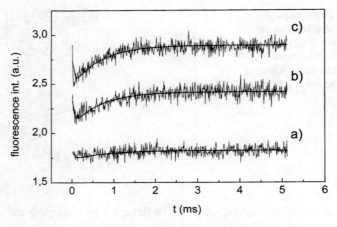

Figure 18. The time evolution of the fluorescence intensity after the application of the Hahn echo sequence for different values of the waiting time τ. The zero of the time axis corresponds to the end of the third microwave pulse. The solid lines are a result of a calculation using Eq. (11). Curve a) corresponds to a τ of 0.2, b) $\tau = 2\,\mu s$ and c) $\tau = 10\,\mu s$.

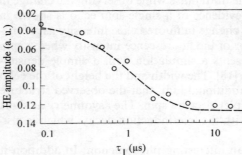

Figure 19. The decay of the Hahn echo intensity as a function of τ, the waiting time between the pulses. The circles represent measured values and the dotted line is a fit with a monoexponential curve.

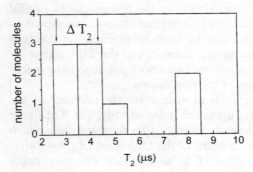

Figure 20. Histogram of the Hahn echo decay times T_2 for 9 different molecules. The width of the boxes is given by the error in the determination of T_2. The two arrows in the diagram represent the calculated variation ΔT_2 of the phase memory time when a variation in the lattice constant of 10% is assumed (see text).

exponential decay function. The decay time of $2\,\mu s$ is within the range known for fully protonated hydrocarbon guest–host systems [25] at low temperatures. The few measured data do not allow for precise determination of the form of the decay curve.

In the following we will discuss the HE decay (spin dephasing) times T_2 for different molecules [37]. The histogram in Fig. 20 compares T_2 for 9 different molecules. The figure shows that the decay times and thus the homogeneous linewidths for different molecules differ by about a factor of three.

It is widely accepted that for localized excited states at low temperatures T ($T < 4\,\text{K}$), the dephasing of the electron spin coherence is caused by the hyperfine interaction of the electron spin and the nuclear spins of the host matrix [25, 38]. The spin configuration of these nuclear spins is a function of time caused by their dipolar spin–spin coupling. These changes in the relative spin configuration mainly take place in form of energy conserving flip–flop processes [251. Those nuclei which have a negligible hyperfine interaction with the electron spin are allowed to participate in this process, whereas those who have a hyperfine interaction with the electron spin much larger than the dipolar coupling to their nuclear neighbors hardly take part in this diffusion and form a so-called "frozen core". Except for the hydrogen nuclei directly bound to pentacene, we will neglect any such frozen core in our calculations since for pentacene in zero magnetic field the hyperfine interaction of the electron spin, even to its nearest p-terphenyl neighbors is very small.

The use of analytical models for spin dephasing in the solid state [39, 40] is not appropriate for the present case because they do not account for the specific distribution of nuclear spins around the electron spin. In order to consider the real matrix structure close to the chromophore we simulated the HE decay curves numerically by using the known crystal structure [9] and the nuclear flip–flop rate W as a parameter. The echo decay is calculated as

$$E(\tau) = \sum_j \cos\left(\sum_i \Delta\omega_i \tau s(\tau)\right) \tag{18}$$

where τ is the waiting time between the pulses, $\Delta\omega_i$ is the change in hfi due to the spin flip of nucleus i and $s(\tau)$ is a step function introduced by Anderson [40] to account for the phase inversion due to the π-pulse. The rate W enters here via the number of changes in $\Delta\omega_i$ during the time interval τ. The first sum runs over all possible nuclear spin configurations j in the surrounding of the electron spin on the chromophore. With a value of $W = 30\,\text{kHz}$ a HE decay time of approximately $3\,\mu\text{s}$ results. To reproduce the dispersion in the decay time T_2 among different molecules, in this model the lattice constant of the surrounding matrix has to be changed, such that the distance between the electron spin and the nuclei in the matrix is modified. For reasons of simplicity we restricted ourselves to the consideration of isotropic changes in the lattice constant. To reproduce a change in decay time T_2 by a factor of two, indicated by the two arrows in Fig. 20, a variation in the lattice constants of 10% has to be assumed [37]. This roughly reproduces experiments by Chan et al. on the pressure dependence of T_2 [45].

We now have to discuss the physical significance of this value. A difference in lattice constants of 10% for the sites of different molecules seem to be surprisingly large at first hand. The frequency position of an optical single molecule transition for example, would shift by several hundred wavenumbers if such a change in the lattice would be produced by pressure [41]. This certainly indicates the simplifications in the model and the above interpretation in terms of a simple change in the lattice constant is only one among a couple of possible explanations. It may be interesting to consider, for example, the effects of defects or vacancies in the host matrix around the chromophore on T_2.

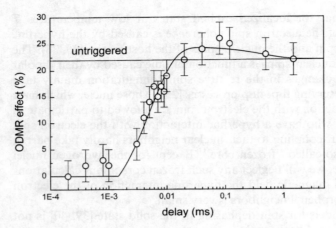

Figure 21. The normalized change of the fluorescence intensity as a function of the delay time between the triggering photon and the application of the microwave pulse. The dots represent measured values and the solid line is the result of a simulation when it is assumed that a single molecule contributes to the fluorescence intensity. The line labeled 'untriggered' gives the ODMR effect for an untriggered application of microwave pulses.

1.6.6.3 Photon-triggered EPR

In analogy to conventional pulsed EPR experiments on photo-excited triplet states it would be advantageous to switch on the microwave field at the moment the triplet state is created. In ensemble experiments this is done by irradiating a microwave pulse at a given delay after a laser pulse has been applied. In these experiments one usually finds a significant EPR signal enhancement. The reason for this is, that at short times after the creation of the triplet state the full polarization of the population is present which develops to thermal equilibrium for longer times. It thus became possible to do conventional microwave-detected EPR on triplet states at room temperature [44]. An equivalent experiment is possible in single molecule EPR at low temperature with continuous laser excitation.

As has been pointed out, the light emission of a single molecule is characterized by the emission of photon bunches interrupted by dark periods when the molecule is in the $^3T_1$ state. Such a behavior is very easily detected for the system terrylene in *p*-terphenyl where k_{23} is so small, that enough photons can be sampled during the singlet interval to observe the quantum jumps to $^3T_1$ in real time [45]. Whenever photons are emitted by the molecule it is in the singlet manifold which is diamagnetic and thus not accessible by the ODMR experiment. The fluorescence intensity can only be affected via microwave irradiation while the molecule is in $^3T_1$. In order to apply a microwave pulse synchronized to the creation of $^3T_1$ one has to trigger the pulse to the detection of a photon. With a suitable delay time one can be sure that the molecule has crossed over to $^3T_1$. A simple method to check whether the experiment works as expected is to record the ODMR effect as a function of the delay time between the triggering photon and the microwave pulse [22]. Such an experiment is shown in Fig. 21. As can be seen the ODMR effect vanishes for short delay times because the microwave pulse is applied when the molecule still is in the singlet manifold. For longer delay times the ODMR effect increases. The time constant for this increase is roughly 30 µs and is determined by the ISC rate k_{23}. The ODMR effect is maximal for a delay time coinciding with the highest probability that the molecule is in $^3T_1$ when the pulse is applied. In this case the triggered ODMR effect even exceeds the ODMR effect when an untriggered train of microwave pulses is applied. This is because in an untriggered experiment a certain separation between the microwave pulses has to be chosen to achieve the maximum ODMR effect. This separation is roughly five times the mean duration of the $^3T_1$ interval [22]. For the triggered experiment the internal "clock" of the molecule determines the separation of the pulses which is, as shown by the calculation (solid line), more effective then the manual adjustment.

1.6.7 Conclusion and outlook

Magnetic resonance on single molecules is the extension of single molecule spectroscopy to the detection of single electron spins and their interaction with the environment. A number of standard methods in magnetic resonance have been applied to single molecules such that in general the field is now open for the application of the versatile instrumentation of magnetic resonance techniques covering the range from extreme spectral resolution to sophisticated double resonance and spin coherent techniques.

Two major spectroscopic topics emerge. Firstly application of magnetic resonance techniques to explore properties of microscopic disorder in relation to static and dynamic parameters in solids. Spectroscopy on single spins will provide additional and more detailed information e.g. on spectral diffusion as has been discussed in previous chapters throughout this book. Secondly it will become possible to determine the full information on interaction tensors even for complete statistical distribution of the orientation of molecules. Besides these intriguing spectroscopic investigations with an extremely high sensitivity basic ideas of quantum mechanics and the interaction of electromagnetic radiation with matter may be illuminated in the future. A first step in this direction are the photon triggered experiments. Further questions might for example be connected with the spinor properties of single hydrogen nuclei.

Since magnetic resonance techniques – in special NMR – are very versatile methods, where are the limitations with respect to single molecule spectroscopy? Probably the most severe limitation is the applicability of single molecule spectroscopy itself, which is presently limited to a small number of chromophores. The magnetic resonance experiment is even more demanding since strict requirements with respect to the population and depopulation rates of the triplet sublevels have to be fulfilled. Presently only the system presented in this chapter has been successfully studied by single molecule ESR. A major issue of future investigations will thus be to increase the number of systems accessible by the technique.

One of the most intriguing features of single molecule detection is the combination with optical microscopy, especially with scanning near-field microscopy as it has been discussed at different places in this book. It can be clearly envisaged that these techniques can be combined with magnetic resonance on single spins. Work is in progress towards this direction. With regard to recent progress in scanning magnetic resonance field gradient microscopy this technique might open the gateway for such futuristic experiments like a magnetic resonance imaging experiment at the level of a single molecule.

1.6.8 Acknowledgement

We would like to acknowledge stimulating discussions with the groups of Dr. M. Orrit (Bordeaux) and Prof. W. E. Moerner (San Diego).

This research has been made possible with financial support from the DFG (SFB 337, Innovationscollege Chemnitz, Contract Bo 935/6-1; Leiden, Contract Ko-1359/ 2-1), the "Stichting voor Fundamenteel Onderzoek der Materie F.O.M.", the Procope program of the DAAD and the HCM program of the EU.

References

[1] C. von Borczyskowski, *Appl. Magn. Res.* **2**, 159, (1991).
[2] R. H. Clarke, *Triplet state ODMR spectroscopy.* Wiley, New York, 1982.
[3] M. Sharnoff, *J. Chem. Phys.* **46**, 3263, (1967).
[4] A. L. Kwiram, *Chem. Phys. Lett.* **1**, 272, (1967).
[5] J. Schmidt and J. H. van der Waals, *Chem. Phys. Lett.* **2**, 640, (1968).
[6] W. G. van Dorp, T. J. Schaafsma, M. Soma, and J. H. van der Waals, *Chem. Phys. Lett.* **21**, 221, (1973).
[7] J. Köhler, J. A. J. M. Disselhorst, M. C. J. M. Donckers, E. J. J. Groenen, J. Schmidt, and W. E. Moerner, *Nature* **363**, 242, (1993).
[8] J. Wrachtrup, C. von Borczyskowski, J. Bernard, M. Orrit, and R. Brown, *Nature* **363**, 244, (1993).
[9] J. L. Baudour, Y. Delugeard, and H. Cailleau, *Acta Cryst.* **B32**, 150, (1976).
[10] J. H. Meyling and D. A. Wiersma, *Chem. Phys. Lett.* **20**, 383, (1973).
[11] T. E. Orlowski and A. H. Zewail, *J. Chem. Phys.* **70**, 1390, (1979).
[12] J. O. Williams, A. C. Jones, and M. J. Davies, *J. Chem. Soc. Faraday Trans. 2* **79**, 263, (1983).
[13] F. G. Patterson, H. W. H. Lee, W. L. Wilson, and M. D. Fayer, *Chem. Phys.* **84**, 51, (1984).
[14] A. J. van Strien and J. Schmidt, *Chem. Phys. Lett.* **70**, 513, (1980).
[15] J. Bernard, L. Fleury, H. Talon, and M. Orrit, *J. Chem. Phys.* **98**, 850, (1993).
[16] W. E. Moerner and T. Basché, *Angew. Chem. Int. Ed. Engl.* **32**, 457, (1993).
[17] L. Kador, D. E. Horne, and W. E. Moerner, *J. Phys. Chem.* **94**, 1237, (1990).
[18] R. Brown, J. Wrachtrup, M. Orrit, J. Bernard, and C. von Borczyskowski, *J. Chem. Phys.* **100**, 7182, (1994).
[19] H. van der Meer, J. A. J. M. Disselhorst, J. Köhler, A. C. J. Brouwer, E. J. J. Groenen, and J. Schmidt *Rev. Sci. Instrum.* **66**, 4853, (1995).
[20] A. C. J. Brouwer, J. Köhler, E. J. J. Groenen, J. Schmidt, and J. Wrachtrup, *to be published.*
[21] C. A. Hutchison, J. V. Nicholas, and G. W. Scott, *J. Chem. Phys.* **53**, 1906, (1970).
[22] J. Wrachtrup, C. von Broczyskowski, J. Bernard, M. Orrit, and R. Brown, *Phys. Rev. Lett.* **71**, 3565, (1993).
[23] J. Köhler, A. C. J. Brouwer, E. J. J. Groenen, and J. Schmidt, *Chem. Phys. Lett.* **250**, 137, (1996).
[24] A. Gruber, M. Vogel, J. Wrachtrup, and C. von Borczyskowski, *Chem. Phys. Lett.* **242**, 465, (1995).
[25] C. A. van t'Hoff and J. Schmidt, *Mol. Phys.* **38**, 309, (1979).
[26] T. Basché, S. Kummer, and C. Bräuchle, *Chem. Phys. Lett.* **225**, 116, (1994).
[27] J. Köhler, A. C. J. Brouwer, E. J. J. Groenen, and J. Schmidt, *Chem. Phys. Lett.* **228**, 47, (1994).
[28] A. C. J. Brouwer, J. Köhler, E. J. J. Groenen, and J. Schmidt, *J. Chem. Phys.* **105**, 2212, (1996).
[29] U. Doberer, H. Port, D. Rund, and W. Tuffensammer, *Mol. Phys.* **49**, 1167, (1983).
[30] T. S. Lin, J.-L. Ong, D. J. Sloop, and H.-L. Yu, In: *Pulsed EPR: a new field of applications*, (Eds. C. P. Keijzers, E. J. Reijerse, and J. Schmidt), North-Holland, Amsterdam, 1989, p. 191.
[31] J. Köhler, A. C. J. Brouwer, E. J. J. Groenen, and J. Schmidt, *Science* **268**, 1457, (1995).
[32] W. G. Breiland, H. C. Brenner, and C. B. Harris, *J. Chem. Phys.* **62**, 3458, (1975).

[33] J. Schmidt and J. H. van der Waals, In: *Time domain electron spin resonance*, (Eds. L. Kevan and R. N. Schwartz), Wiley, New York, 1979, p. 343.

[34] W. G. Breliand, C. B. Harris, and A. Pines, *Phys. Rev. Lett.* **30**, 158, (1973).

[35] R. Kettner, J. Tittel, T. Basché, and C. Bräuchle, *J. Phys. Chem.* **98**, 6671, (1994).

[36] B. Kozankiewicz, J. Bernard, and M. Orrit, *J. Chem. Phys.* **101**, 9377, (1994).

[37] J. Wrachtrup, C. von Borczyskowski, J. Bernard, R. Brown, and M. Orrit, *Chem. Phys. Lett.* **245**, 262, (1995).

[38] J. Schmidt, In: *Relaxation processes in molecular excited states*, (Ed. J. Fünfschilling), Kluwer, Dordrecht, 1989, p. 3.

[39] P. Hu and S. R. Hartmann, *Phys. Rev. B* **9**, 1, (1973).

[40] P. W. Anderson, B. I. Halperin, and C. M. Varma, *Philos. Mag.* **25**, 1, (1972).

[41] M. Croci, H. J. Müschenborn, F. Güttler, A. Renn, and U. P. Wild, *Chem. Phys. Lett.* **212**, 71, (1993).

[42] P. F. Jones, *J. Chem. Phys.* **48**, 5448, (1968).

[43] I. Y. Chan and X. Q. Qian, *J. Chem. Phys.* **95**, 7076, (1991).

[44] D. Stehlik, C. H. Bock, and M. C. Thurnauer, In: *Advanced EPR in biology and biochemistry*, (Ed. A. J. Hoff), Elsevier, Amsterdam, 1989.

[45] T. Basché, S. Kummer, and C. Bräuchle, *Nature* **373**, 132, (1995).

[46] H. de Vries and D. A. Wiersma, *J. Chem. Phys.* **70**, 5807, (1979).

[47] M. Orrit and J. Bernard, *Phys. Rev. Lett.* **65**, 2716, (1990).

2 Near-field Optical Imaging and Spectroscopy of Single Molecules

2.1 Introduction

The unifying theme of this chapter is a method, namely near-field optical excitation. We will, therefore, not only review the single molecule data obtained using near-field optics but also discuss the method, in particular its advantages and disadvantages vis-à-vis conventional 'far-field' optics. Historically, single molecule spectroscopic studies were first performed at low temperatures using conventional optics [see Chapter 1] and at room temperature using near-field optics. There has, consequently, been a tendency to associate far-field with low-temperature and near-field with room-temperature studies. There is nothing fundamental about this association. A number of the low-temperature experiments have now been carried out using near-field, and conversely, room-temperature experiments using far-field are proliferating. Certainly one of the primary purposes of compiling a book such as this is that it might serve as a resource for researchers seeking to enter the field. In this spirit, we will review the basic near-field optical technology, examine the fundamental and practical differences between near- and far-field optics as pertinent to the study of single molecules, and then review the single molecule data acquired to date using near-field optical techniques.

Before beginning, a few comments regarding the term 'molecule' are appropriate. Consider, for example, the DNA in a single chromosome. When isolated and stretched out, one strand can be several centimeters long. This is a single molecule. A large number of spatially separated and independently fluorescing moieties may be attached to the DNA to become part of this large, single molecule. Each of these independent moieties are single fluorophores, but they are not single molecules because they are part of the larger whole. It is the single fluorophores that are of interest here. The term 'single-molecule spectroscopy' should be read implicitly as 'single-fluorophore spectroscopy'.

2.2 Principles

2.2.1 Near-field optics

The near-field idea is simple and elegant but rather difficult to implement. This accounts for its repeated and apparently independent 'discovery'. Had it been readily demonstrated, the idea would not have been lost after it was proposed in 1928 by Synge or again by O'Keefe in 1956 [1]. An experimental realization came, finally, in 1972 with Ash and Nicholls $\lambda/60$-resolution images utilizing microwave radiation of wavelength $\lambda = 3$ cm [2]. In classical optics, the resolution of an image is constrained to $\approx \lambda/2$ [3]. This result holds for systems where the image is formed at a distance of many wavelengths from the imaging optic. If, however, one can create a localized source of radiation having a size much smaller than the wavelength and place it within a fraction of a wavelength of the surface of interest, then the area illuminated will be roughly the size of the source. Given such a source, an image of the surface can then be created, as in any scanning microscope, by rastering the source relative to the sample to build up an image pixel by pixel. The data presented in this chapter were obtained using such a near-field radiation source. Synge's original proposal consisted of a small aperture in an opaque screen. The modern embodiment of this idea is the metal-coated tapered optical fiber probe [4]. As, by definition, the near-field source must be close to the surface, it is advantageous to have the opaque screen bend back away from the surface. Further, it is imperative that the radiation be delivered to the aperture as efficiently as possible. Finally, it is preferable that the means of localizing the radiation field not interfere with the otherwise desirable properties of that field. For example, when performing a spectroscopic study one may want a narrow-linewidth tunable source or, conversely, a short-pulse source. The tapered optical fiber has proven to be the best general solution to these criteria, to date.

The above description of the near-field concept is qualitative. A somewhat more detailed description is necessary in preparation for a thorough examination of the experimental data and the differences between the near-field and far-field methods. Toward this end, consider a monochromatic electromagnetic radiator having dimensions much smaller than the wavelength. Far from the source, radiation propagates with wavevector $k = 2\pi n/\lambda$, where n is the index of refraction of the surrounding medium. Near the radiator, there are components of the optical field having $k > 2\pi n/\lambda$ that cannot propagate away from it into the far zone. This additional power circulates in the near-field of the radiator. Matter within the near-field region interacts with the total field including the non-propagating components. Since the total field falls off rapidly from the aperture, there is an enhanced sensitivity to the matter near the aperture. Thus, the high-resolution capability of near-field microscopy/spectroscopy depends on both the lateral confinement of the field and its limited 'depth-of-field'.

2.2.2 Sensitivity

The sensitivity of a fluorescence experiment depends not only on the collection and detection efficiency of the apparatus but on the relative magnitude of the background light. Suppose the sample generates a background due to inelastic scattering and/or weak luminescence, then the smaller the sample volume, the smaller the background. Thus, it is easy to understand the basis of the argument that the signal-to-background ought to be superior in near-field compared to far-field experiments: Using a near-field probe capable of $\lambda/40$-resolution [5], the sample volume is some four orders of magnitude smaller than that of a diffraction-limited lens. Whereas there is nothing fundamentally wrong with this argument, there are complications that adversely affect the signal-to-background achievable in present state-of-the-art near-field experiments. First, there is the unavoidable silica Raman from the optical fiber probe itself. In a typical experiment using a ≈ 1 m-length of fiber, the number of counts in the peak of the Si–O stretch band at ≈ 550 cm^{-1} is comparable to the number of counts at the maximum of the fluorescence emission from a single molecule (at room temperature). Second, there is a diffuse luminescence emission emanating from the aperture of the fiber probe in the 'yellow' spectral region, the relative intensity of which appears to increase with smaller apertures. Finally, and most importantly, single-molecule experiments have not yet been performed at the highest resolution achievable with near-field but rather at a resolution between $\approx \lambda/10$ and $\approx \lambda/2$. The primary reasons for this are the inherently weak interaction of a room-temperature single molecule with the radiation field and the consequent requirement of a relatively high field, and the temperature sensitivity of the sample. The connection between these properties and the probe size requires a short digression on the properties of the Al-coated, tapered optical fiber probe.

In the vicinity of the aperture, the incident light impinges on the probe's aluminum coating. The majority of the field is excluded from the metal but some is absorbed. For reference, the reflectivity of an aluminum mirror at normal incidence is only $\approx 92\%$ in the mid visible. The absorbed light heats the metal. The extreme proximity of the tip and sample results in a local sample temperature approaching that of the tip. Consequently, the light flux must be maintained below a level that would raise the probe temperature substantially above ambient. On the other hand, the acquisition of the fluorescence emission spectrum of a single molecule in a reasonable time (say, one minute) requires a certain flux. The connection between the tip temperature and the energy density at the aperture is the highly nonlinear transmission of the fiber probe as a function of aperture size. In the simplest model, due to Bethe [6], the transmission goes as a^6, where a is the aperture radius. Recent numerical simulations indicate that the transmission function of the aluminum-coated, tapered fiber probe has an even higher power dependence [7]. Without delving further into the details, suffice it to say that to achieve a photon flux of 10^{20}–10^{21} cm^{-2} s^{-1} without heating the tip, it is necessary to use aperture diameters no smaller than $\sim \lambda/5$, with the present generation of probes.

The upshot of all this is that the kind of experiments performed on single molecules using near-field to date could have been done just as well with diffraction-limited excitation. There remain, however, the studies which are no doubt to come where the higher resolution of near-field will be necessary. It is in this realm that near-field should find its niche.

2.3 Experimental methods

2.3.1 Fluorescence excitation and detection

Fig. 1 is a stylized depiction of a near-field probe tip and the optical electric field merging from the aperture. A probe is fabricated by first heating and pulling a length

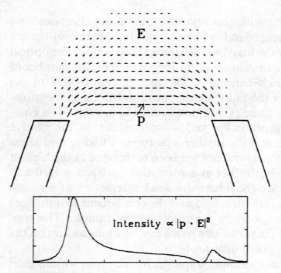

Figure 1. (Top) Schematic view of a molecular transition dipole *p* at a particular orientation within the electric field pattern *E*, of a subwavelength aperture. (Bottom) Resulting excitation rate *vs.* lateral displacement for this particular orientation which is proportional to the square of the component of *p* along *E*. Adapted from Ref. 13.

of a single-mode optical fiber so as to form an adiabatically-tapered dielectric waveguide. Then the tapered fiber is metal coated *via* a shadow evaporation procedure so that the end face is left as a clear aperture in the metal film. The process is best documented in the paper by Valaskovic *et al.* [8]. One point that is frequently overlooked is the importance of efficiently guiding the light to the aperture and the sensitivity of the throughput to the form of the taper and to the initial mode distribution in the fiber. The latter issue is optimized by the use of single-mode fibers; the former is still something of an art. The importance of the taper was emphasized in Ref. 8 with examples of transmission efficiencies varying by 1–2 orders of magnitude for fibers having identical aperture diameters. Consequently, it is important to note not only the aperture size but also the transmission coefficient in assessing the quality of a probe. The transmission coefficient is a somewhat arbitrary quantity formed by the ratio of the transmitted intensity collected by a high numerical aperture (NA) objective to the intensity in the core of the single-mode fiber. Numbers for this quantity have ranged from $\approx 10^{-4}$ to 10^{-5} in the single molecule experiments. The coating is usually $\approx 100\,\mathrm{nm}$ of aluminum, the quality of which will determine the closeness of approach of the effective aperture to the sample.

Fig. 2 is a schematic of the experimental setup of one of the authors. Although the particulars may vary, the elements of this apparatus are included in most, if not all, the room-temperature single-molecule near-field scanning optical microscopes. The probe aperture is held in close proximity to the sample surface. A typical sample consists of a sparse coverage of fluorescent molecules either directly deposited on fused silica or in/on a thin, ≈ 20–$30\,\mathrm{nm}$, layer of polymer on fused silica. Fluorescence from the molecules is collected through the substrate using a high numerical aperture, NA ≈ 1.2 to 1.4, oil-immersion objective. The half-angle of collection of such an objective is ≈ 60–$70°$, and the corresponding solid angle fraction is $\approx 1/4$. The fraction of the total molecular fluorescence emission collected by the objective is not,

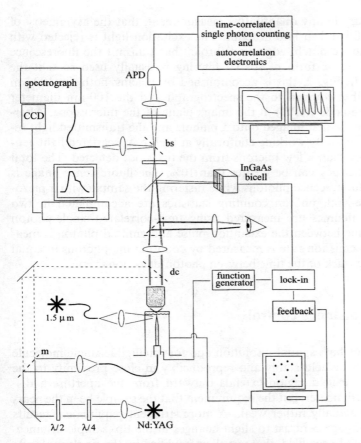

Figure 2. Schematic of the near-field/far-field time-resolved imaging spectroscopy apparatus. The pulse train from a Nd : YAG (Coherent Antares) is directed into an optical fiber probe or the back of an objective for near- or far-field excitation, respectively, depending on the positions of the removable mirror, 'm', and dichroic beamsplitter, 'dc'. Fluorescence emission collected by the immersion objective (*NA* 1.25–1.40) is imaged onto an avalanche photodiode, 'APD', and/or a multimode fiber coupled to a spectrometer, depending on the position and coating of the beamsplitter, 'bs'. Probe-sample distance regulation is accomplished through a constant-dither-amplitude feedback loop, the dither amplitude being measured by imaging the 1.5-μm light scattered off the probe onto an InGaAs bicell. Scanning, lifetime measurement and spectroscopy are controlled from three separate computers, as shown. Adapted from Ref. 18.

however, given by the simple geometric fraction of the total solid angle subtended by the objective. In the experimental arrangement just described the molecule exists at a dielectric interface. The dipole radiation pattern of the molecule is perturbed by the presence of the interface so that most of the emitted radiation goes into the half space defined by the high refractive index medium [9]. In the near-field experiment, the issue is further complicated by the presence of the probe, as will be discussed further below. In any case, one must be careful when assessing the actual collection efficiency, as is necessary in any quantitative measurement, that the asymmetry of the molecular emission is taken into account. The excitation light is rejected with filters, such as holographic notch, edge or dichroic, that transmit the fluorescence emission wavelengths. Some form of spatial filtering is usually used so that the detection is confocal. In Fig. 2, this is accomplished by placing both the 100-μm diameter multimode fiber coupled to the spectrograph and the 100-μm diameter silicon avalanche photodiode (APD) in the image plane of the microscope. Alternatively, the sample can be first imaged onto a pinhole and the transmitted light re-imaged onto the detector(s). By working confocally at high NA, scattered light generated at distances more than a few microns from the tip is not detected. The total collection/detection efficiency can be greater than 10%. The fluorescence image is recorded by counting fluorescence photons while rastering the sample with a piezo-electric scanner. In general, photon counting statistics are accumulated in two modes. Excited state lifetimes are measured using time-correlated single-photon counting, where the time between the excitation pulse and emitted photon is measured. In addition, the emission rate is measured by counting the photons in equal time bins or by keeping track of the time between photon-detection events.

2.3.2 Tip–sample distance control

It is paramount both for high spatial resolution and for intelligible single molecule results that the aperture be held stably and reproducibly in close proximity to the sample surface. The near-field region extends outward from the aperture a distance on the order of its diameter, but the requirement that the sample be in the near-field of the aperture is actually rather weak. A more stringent requirement results from the sensitivity of image contrast to slight changes in the tip–sample distance. This is a general feature of near-field, that is well exemplified by the single molecule lifetime measurements discussed below. Two methods have been used to sense and control the tip-to-sample gap, shear-force microscopy (SFM) [10] and background fluorescence increase (BFI) [11].

The principle behind SFM is that the lateral or shear force between an oscillating probe tip and the sample increases as the distance decreases. The probe is usually mounted in a support such that several millimeters of the aperture end of the optical fiber extends beyond the clamping point. The probe thus forms a cantilever having one fixed and one free end. It is driven transversely at a so-called 'tip resonance', which indicates that the resonance is due to the cantilever rather than the support structure of the microscope, with an amplitude ≈5 nm. Shear forces between the probe tip and sample surface damp the oscillation. The amplitude is measured and fed back to the sample height position so as to maintain constant oscillation amplitude and presumably constant tip–sample distance. The amplitude was measured, originally, with optical deflection methods. Recently, a number of electrical measurement schemes have been demonstrated that may prove to have a number of advantages in speed, sensitivity or 'ease-of-use' [12]. In near-field single molecule experiments the bandwidth of the feedback is not an issue as scan rate is limited by the photon count rate to ≈ 10 ms per pixel. With an integration time of 10 ms, the RMS height noise in NSOM is typically less than ~ 1 nm.

Background fluorescence increase is based on a process that is related to the inverse of total internal reflection (TIR). In TIR, light incident on a dielectric interface from within a higher-index medium at an angle larger than a critical angle is totally reflected. An evanescent field exists on the lower-index side that, by definition, decays exponentially away from the surface with a characteristic length on the order of the wavelength. The inverse of this process occurs when an optical radiator in the lower-index half-space is brought within an evanescent decay length of the higher-index medium. Large wavevector components of the optical near-field of the radiator are coupled into the higher-index medium at angles larger than the critical angle. This additional "forbidden light" results in increased transmission into the substrate and increased excitation of background fluorescence from out-of-focus molecules. Forbidden light position sensing was used in low temperature near-field excitation of single molecules [11].

A few caveats about probe–sample distance should be made. In SFM, the oscillation amplitude falls off over a range of <50 nm, and often <10 nm. Zero amplitude does not necessarily correspond to 'contact'. It can correspond to a position some distance above hard contact if the tip is in a liquid or contamination layer. Values of 'tip height' are, strictly speaking, extrapolated from the position of zero amplitude. Stated distances from the surface should be viewed as a level of damping. It might be helpful for authors to state the starting amplitude as well as the operating amplitude.

2.4 Results

2.4.1 Systems studied to date

Table 1 lists the samples wherein single molecules have been detected using near-field. Generally, the substrate is a silica coverslip, sometimes coated with a thin layer of polymer, sometimes not. In the former case, the polymer is spin-coated onto the silica from a dilute solution resulting in a continuous film ~ 10–30 nm thick. Dye molecules are deposited on the surface by allowing a drop of a dye solution to dry, by spin-coating a drop of the solution onto the surface, or by spin-coating a mixed solution of dye and polymer onto the substrate. The preparation of the sample for the low temperature work is covered in Section 1.1.

2.4.2 Orientation and location

The first report of single molecule detection in an optical microscopy under ambient conditions was that of Betzig and Chichester [13]. The most striking aspect of their single molecule fluorescence images, reproduced in Fig. 3(a), is the variety of shapes. Except for the occasional photobleaching event which resulted in the disappearance

Table 1. Single molecule system studied with near-field excitation.

Molecule	Host	Reference
DiI	PMMA on silica	13, 17, 18
Rhodamine 6G	silica	21, 27
Rhodamine 6G	PVB on silica	30
Sulfarhodamine 101	silica	24, 25
Tetramethylrhodamine, Texas Red	covalently linked to DNA, dried onto silca	33
Pentacene	*p*-terphenyl	11
Rhodamine-phalloidin-labeled cytoskeletal actin*	cell, dried on silica	34
allophycocyanin proteins*	silica	36
B-phycoerythrin proteins*	silica	29
TOTO-l-stained DNA*	silica	29

* In these experiments, the fluorescence originating from a few molecules was measured.

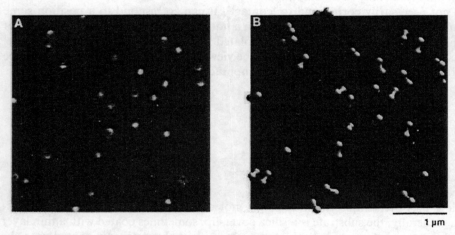

1 μm

Figure 3. (A) $4 \times 4\,\mu m^2$ image of a field of single molecules obtained under random polarization. (B) Stylized map indicating the dipole orientations of molecules in (A) as inferred from polarization data and the model represented in Fig. 4, below. Each dipole is represented as a dumbbell centered at the molecular position, with a length determined by the projection of the dipole onto the *x, y* plane, and an angular orientation determined by ϕ. Adapted from Ref. 13.

of a molecule, the molecular images were largely invariant upon repeated scanning. Yet, the images were strongly dependent on the polarization of the excitation and/or emission fields. By referring back to Fig. 1, one can easily understand these results. A molecule interacts with an oscillating electric field via its transition dipole moment, **p**. The spatial range of this interaction is on the order of $\sigma^{1/2} \approx 0.1$ nm for dye molecules at room temperature, where σ is the absorption cross-section. Thus, a molecule probes the electric field on a length scale much smaller than that over which the field amplitude varies, even when the field is highly confined as in the vicinity of the probe

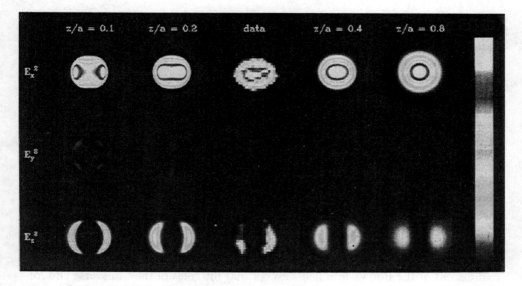

Figure 4. Squared components of the electric field predicted to exist at various normalized distances (z/a) from a subwavelength diameter aperture (calculation after Bouwkamp, Ref. 14), as compared to actual single molecule data (center). Adapted from Ref. 13.

aperture. As the molecule is scanned across the aperture, the absorption probability and, consequently, the fluorescence intensity varies in proportion to $|\boldsymbol{p} \cdot \boldsymbol{E}|^2$. Based on these facts, the interpretation of the variously shaped fluorescence images in fig. 3a is that each is an image of the electric field intensity along an axis determined by the orientation of the absorption transition dipole of a single molecule. The variety of image shapes is due to the distribution of dipole orientations.

Substantiation for this interpretation is provided in Fig. 4 where a number of plots of the electric field intensity in the vicinity of an aperture are presented along with two plots of single molecule data. Betzig took as a model for the near-field of the fiber probe the field calculated by Bouwkamp [14] for the case of an aperture in a perfectly conducting, infinitesimally thick metal film illuminated from the back side by a monochromatic plane wave. Whereas this model cannot be quantitatively correct, it is clearly correct, qualitatively. The greatest deviation between the true field and the Bouwkamp field will occur close to the metal edges where singularities exist for a perfect metal. Recently, numerical calculations have been performed for the case of a more realistic, yet still idealized, tip model [15]. The main features of any of these calculations are that the electric field is normal to the metal surfaces at the aperture edges and normal to the probe axis in the center. Therefore, molecules having $\boldsymbol{p}$ normal to the sample plane are most strongly excited by the edge fields resulting in a double arc image, and molecules with $\boldsymbol{p}$ in the sample plane produce elliptical shapes. Betzig used the shapes of the single molecule fluorescence images under two excitation polarization conditions to determine the orientation of the absorption transition dipole as shown in Fig. 3(b). Note that because the field shape changes as a function of distance from the aperture, in principle the distance of the molecule in the direction normal to the aperture plane can be determined as well.

To conclude this section, a number of comments are in order. First, the preceding discussion has ignored the possibility that the fluorescence quantum yield is a function of the relative positions of the probe and molecule which, as will be shown below in the section on lifetime measurements, is not strictly true. Second, the different shapes discussed above are not always observed. Assuming a sample having a distribution of molecular orientations, including some having an out-of-plane (z) component, then the failure to observe fluorescence images indicative of a normally oriented ('z') molecule can only be because the probe used does not produce a Bouwkamp field. Factors that affect the field distribution at the aperture are tip morphology, i.e., the lumpiness of the aluminum coating, the shape of the tip, and the aperture size. The ability to resolve 'z's' is a convenient measure of probe quality. Finally, as recently demonstrated by Macklin and one of the authors [16], one can create a longitudinal electric field having a magnitude comparable to that of the transverse field using standard diffraction-limited optics. The field distributions are reminiscent of the Bouwkamp fields except that the roles of the transverse and longitudinal components are reversed. One means of doing this is in an optical fiber. It is conceivable that a fiber probe having an aperture diameter larger than the cutoff of the TM_{01}, mode would have a double-lobed transverse field distribution and a longitudinal field centered on the probe axis.

In the first low temperature single molecule experiments using near-field excitation, Moerner, Plakhotnik, Irngartinger, Wild, Pohl and Hecht [11] used the spatial distribution of the field and the field gradient near the probe aperture to obtain the 3-dimensional positions of molecules below the surface of a crystal. As discussed in Section 1.1 of this book, the zero-phonon lines (ZPL) of guest molecules in low-temperature matrices are extremely narrow and, consequently, have very large absorption cross-sections. This results in high sensitivity to external influences, measurable as small ($\delta v/v \approx 10^{-7}$) changes in the width and center frequency of the single molecule ZPL. In the Moerner experiment, the molecules were pentacene in a p-terphenyl crystalline host. As is the case in far-field low-temperature single-molecule experiments, a large number of pentacene molecules were present in the illuminated volume. Fig. 5 shows the statistical fine structure that results from tuning the single frequency laser through a portion of the inhomogeneously broadened ZPL and its variation as the sample is moved into the near field of the probe. Because the illuminated volume is so much smaller when using near field excitation, single molecule features are apparent in Fig. 5 even though the sample is high concentration and the laser is tuned near the inhomogeneous line center. Moerner *et al.* further used the high field gradients near the probe to obtain information on the relative depths of the single pentacene molecules, as shown in Fig. 6. First, molecules close to the probe could be discriminated against those further away by investigating optical saturation of the fluorescence excitation signal. The fluorescence from molecules close to the probe plateaus at lower powers than those further away. Second, by applying a static voltage to the probe ($\approx 10\,V$), optical resonances were shifted due to the quadratic

Figure 5. Statistical fine structure and single-molecule features upon approach to the near-field. Approximate distances from the surface: (a) 1.2 μm, (b) 0.5 μm, and (c) 210–270 nm. Each panel shows two spectra taken about 5 min apart to show reproducibility. Zero detuning = 592.066 nm, the tip voltage, $V_T = 0$, and all traces are from the same sample. Adapted from Ref. 11.

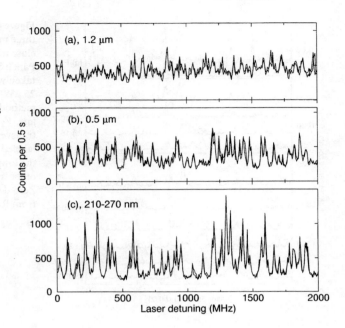

Stark effect. Molecules that displayed low power saturation also showed the largest Stark shifts, again indicating their close proximity to the probe. By examining the magnitude of the Stark shift, an estimate of the distance of the molecule from the tip was made, with an accuracy of about 100 nm. Finally, the tip was oscillated laterally with an amplitude of 20 nm creating a modulation of the Stark shift which could be used to determine the polar angle, θ, of the molecule to the tip axis.

2.4.3 Spectral diffusion at room temperature

Another effect commonly observed in far-field low-temperature single molecule spectroscopy is spectral shifting due to effects other than macroscopically applied fields. Some shifts, termed spectral diffusion, are spontaneous and others, such as those responsible for hole burning, are photo-induced. At liquid He temperatures, spectral diffusion due to local physical perturbations has been observed to occur in narrow frequency intervals (<GHz). As the temperature is raised from a few Kelvin to room temperature, the rate of diffusion and the spectral linewidth increase dramatically, which tends to wash out these effects on time scales accessible to measurement. So it was somewhat of a surprise when spectral shifts in emission spectra were observed at room temperature.

In the first room-temperature, spectrally-resolved, single-molecule experiments, Trautman, Macklin and coworkers observed spectral shifting [17]. The sample consisted of diI (an indocarbocyanine dye) dispersed on polymethylmethacrylate

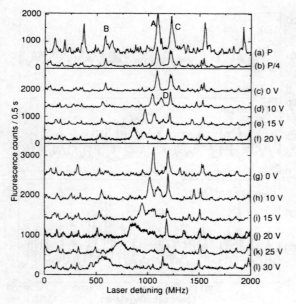

Figure 6. Excitation spectra showing three methods to identify molecules close to the tip with $d \approx 250$ nm. Upper Panel: Saturation method; spectra taken with (a) $P_C = 100$ μW, and (b) 25 μW. Middle Panel: Static Stark shift method, traces labeled by the tip voltage, V_T. Bottom Panel: Stark shift with transverse dithering method, traces labeled by V_T. The scale is exact for the lowest trace in each panel with the other traces shifted vertically upward. Zero detuning = 592.067 nm. Adapted from Ref. 11.

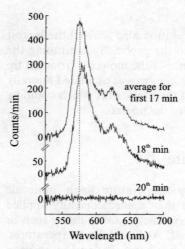

Figure 7. Single-molecule spectra of a highly photostable molecule. The upper trace is the average of 17 spectra acquired over 17 min during which no distinguishable spectral change occurred. During the eighteenth one-minute integration, the emission red-shifted by about 5 nm. In the nineteenth minute, the molecule photobleached. The lowest trace (labeled twentieth minute) is the background. It was calculated that approximately 2×10^8 photons were emitted by this molecule during the spectral integration. Adapted from Ref. 17.

(PMMA). In Fig. 7 an example of such a shift is displayed. A concern with a new technique such as near-field spectroscopy is that the measurement might perturb the process under observation. Here, the issue is whether the spectral variability is induced by the probe. Consider the data presented in Fig. 8. The fluorescence images of two molecules, lying some 150 nm apart are shown with the corresponding emission spectra. The spectra are similar in shape and width but displaced by ≈ 10 nm. In this case, the spectra were invariant during the entire course of the measurement. An additional type of spectral variability is shown in Fig. 9. Taken together, these data

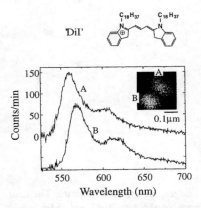

Figure 8. The near-field fluorescence image of two adjacent molecules A and B (see inset) and the corresponding fluorescence spectrum of each. The data were obtained as follows. First, the image was acquired, then the fiber probe was moved over molecule B where it was held stationary while collecting, dispersing and integrating the fluorescence in 1-min. intervals until photochemical degradation extinguished the emission. The resulting spectrum is shown as curve B. The sample area was then re-imaged revealing only molecule A, and the fiber probe was then moved over molecule A where the spectrum labeled A was measured in the same manner. The displayed spectrum B is the average of 21 one-minute integrations, and lies near the center of the distribution of fluorescence wavelengths. Spectrum A is the average of five one-minute integrations, and lies on the high-energy side of the distribution, being shifted about 10 nm to the blue. Each 'count' on the λ-axis of the spectrum is the result of ca. 200 photons emitted by the molecule. Spectrum B is displayed vertically for clarity. The molecular structure of the molecule, DiI, is shown at the top of the figure. Adapted from Ref. 17.

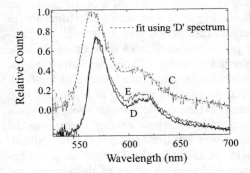

Figure 9. Spectra of two molecules, D and E, exhibiting different intensities in the vibronic band, and of one, C, having a broadened spectrum. The fit to the spectrum of molecule C is the convolution of D with a 15-nm gaussian (FWHM), shifted to shorter wavelength by 5 nm. Although molecule C photobleached after a single 1-min. integration, photostability and spectral width were generally uncorrelated. E was shifted 5.5 nm to longer wavelength to facilitate bandshape comparison, and D and E were displaced vertically for clarity. Adapted from Ref. 17.

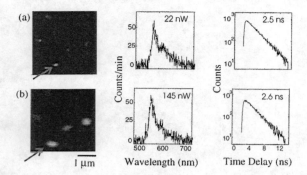

Figure 10. Comparison of near- and far-field fluorescence images, spectra and lifetimes. The near-field images of the two molecules in the upper left quadrant of the scan field are indicative of '*z*'-oriented transition dipoles; the far-field images of those two molecules are merely weak. The spectrum and lifetime of the molecule indicated by the arrow were taken simultaneously under (a) near-field and (b) far-field excitation, respectively. The corresponding excitation powers are included in the figure. Adapted from Ref. 18.

illustrate the range of phenomena observed. First, the ensemble spectrum was in-homogeneously broadened, the spectra of individual molecules exhibiting shifts of ± 8 nm relative to the average spectrum. Second, the single-molecule spectra varied in both shape and width. The former is due to redistribution of the Frank–Condon intensities and is indicative of a symmetry-breaking, local interaction. The latter was interpreted in terms of local motions of the matrix molecules. From polarized mea-surements, it had been determined that the diI molecules were not rotationally diffus-ing on the timescale of the measurement. Third, approximately 20% of the molecules studied exhibited a spectral hop, such as that in Fig. 7. A minimalist interpretation of all these phenomena focused on the dispersion interaction and its spatial and tem-poral variation. Returning to the issue of perturbation, checks were made that the single-molecule spectra summed to give the many-molecule spectrum and that the distribution of photobleaching probabilities was the same near-field and far-field. This was only true, however, when the input power was reduced to a fraction of a mW. Recently, with the advent of room-temperature far-field single molecule spec-troscopy, the measurements have been repeated in a manner exemplified in Fig. 10 [18]. In this experiment time-resolved spectra were obtained from the same individ-ual molecules, both near-field and far-field, confirming the absence of a probe-induced spectral perturbation. One issue not addressed in either of the above cited papers was whether the spectral jumps are photoinduced. The Battelle group is presently examining this question on a variety of single-molecule samples utilizing diffraction-limited excitation [19]. Finally, the issue of resolution, not being central to photophysics, has not been mentioned, yet it remains the singular advantage of near-field. Referring to the data in Fig. 8: With the probe held stationary over the single molecule, B, there was no measurable excitation of molecule A which lay approximately 150 nm away.

2.4.4 Fluorescence lifetime behavior

It has been known for almost thirty years that molecular emission is altered near an interface [20]. Experiments performed in late 60's by Drexhage and Kuhn used Langmuir–Blodget films as spacer layers to place fluorophores at fixed distances from plane metal surfaces. Two physical effects were identified. If an object having a non-zero imaginary part of its dielectric function at a frequency ω is brought within the near-field, $\approx 10\,nm$, of an optical dipole radiating at frequency ω, then direct energy transfer can occur via dipole–dipole coupling. Molecular spectroscopists will be familiar with this concept as Förster energy transfer which is a term that is normally reserved for the case of resonant coupling between two molecules. At greater separations, extending out to distances comparable with λ, the effect of the interface can be described in term of the Fresnel coefficients, that is, in terms of the usual reflection, transmission and absorption coefficients. The classical description given to the effect of the reflected field on the emission is that the reflected wave interferes with the outgoing wave either constructively to enhance emission or destructively to suppress emission. In this regime, the fluorescence lifetime, τ_f, oscillates as a function of the relative separation of the molecule and the surface. To put it in the best possible light: It was suspected that *interesting* fluorescence lifetime behavior would occur close to a near-field probe. To put it in the context of the preceding section, there was a concern that the probe would perturb the lifetime.

To obtain fluorescence lifetimes time correlated single photon counting (TCSPC) was used. In TCSPC, the elapsed time is measured between an excitation pulse from a pulsed laser and a detected photon. A histogram of the elapsed times provides a fluorescence decay curve, from which the fluorescence lifetime, τ_f, is extracted. Examples of decay curves for bare silica and single R6G molecules on silica taken with a near-field probe are shown in Fig. 11 [21].

Pulsed excitation with time-gated detection can lead to reduction of background and improved signal to noise ratio (S/N), as shown in Fig. 11. Ungated histograms for a typical R6G molecule (A) and the bare silica substrate (B) at a distance $<10\,nm$ from the probe are shown in the top panel. The peak in traces (A) and (B) is primarily Raman scattering generated in the 30 cm optical fiber of the probe. After the peak, (B) shows a weak tail of constant background. The tail in (A) above the background is fluorescence from a single R6G molecule. For rejection of the prompt background in emission rate measurements, a time window is opened from 0.7 to 10 ns after the peak, and only photons satisfying this gate are counted. Time gating decreases the integrated background by an order of magnitude and the noise in the background by a factor of 3.

For R6G on silica, experiments were performed on populations of molecules to obtain the mean, unperturbed lifetime, τ_f, and its distribution. Five unperturbed fluorescence decay curves were measured for tips positioned ~ 1.0 to $1.1\,\mu m$ above high-coverage surfaces ($\approx 10^3$ to 10^4 molecules illuminated) (Fig. 11(D)). The decay curves were fitted to a single exponential with $\tau_f = 3.65 \pm 0.04\,ns$, which is in good agreement with the $3.5 \pm 0.1\,ns$ obtained previously [22]. Statistical noise and instrument nonlinearity place an upper limit on a possible Gaussian standard devia-

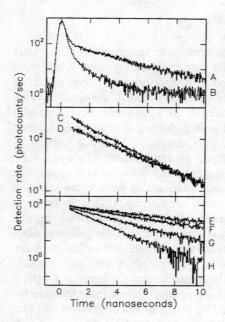

Figure 11. Time-correlated single-photon counting rate histograms. Top Panel: Plots obtained from a single Rhodamine-6G molecule on silica (trace A) and a bare silica substrate (background) <10 nm under an NSOM probe (trace B). Center Panel: Decays obtained for a higher coverage surface with the tip at a distance of <10 nm (trace C) and 1.1 micrometers (trace D) above the surface. Bottom Panel: Traces obtained at lateral positions of maximum fluorescence intensity ($\pm$10 nm) over different single Rhodamine-6G molecules. The extracted lifetime values are 4.6 (trace E), 3.4 (trace F), 2.1 (trace G) and 1.3 ns (trace H). The rates are normalized to 10^3 per second at zero time for slope comparisons. Background traces were subtracted in the center and bottom panels. Farfield powers: (Top) 33 nW, (Center) 1.8 nW, and (Bottom) $\approx$20 nW. The bin width was 30.8 ps. Adapted from Ref. 21.

tion of τ_f values, σ, of $\sigma/\tau_f = 0.2$, which is consistent with theoretical limits on σ for single decay data with noise [23]. A high coverage surface was moved into the near field (<10 nm from the probe face) (Fig. 11(c)). Approximately 25 molecules were distributed in the near-field, and the decay is more nonlinear (on a semilog plot) than the bulk decay. The far-field, bulk decay curves could be consistent with an inhomogeneous distribution of lifetimes with width <20% of the mean. This width increases in the near field.

Using a low-coverage surface, single R6G molecules were positioned within the near field at places of maximum fluorescence intensity, and data were collected until each molecule photobleached (yielding from 10^3 to 10^5 photocounts). Examples of fluorescence decay curves from individual R6G molecules are shown in Figs. 11(e–h). Each is well fitted by a single exponential, but they are all very different from the bulk decay. This occurs because the molecules had different orientations, and the position of maximum fluorescence corresponded to different positions under the tip.

Single molecules were moved laterally across the face of the near-field probe, and the emission rate and τ_f were measured at fixed sample displacements (Fig. 12). The data were obtained at power levels differing by a factor of 8 to 9. The emission rates scale approximately with the power, indicating that the higher irradiance is below the saturation value and that tip heating effects do not alter the fluorescence yield. Fluorescence decays were obtained concurrently with the emission rate data from which single exponential lifetimes were determined. These data were taken with the shear-force signal set at <10 nm from the position of zero amplitude. Near the metal coating at the edges of the aperture ($\sim \pm$150 nm), τ_f is short, and as a molecule

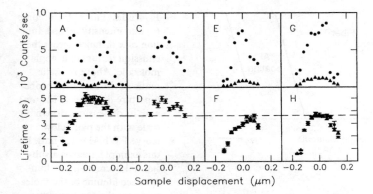

Figure 12. Emission rate (Top Panels) and Fluorescence lifetime (Bottom Panels) versus lateral displacement for four different Rhodamine-6G molecules. The data in panels A and B are from the molecule labeled A in Fig. 22. The tip was maintained <10 nm above the surface throughout these measurements (the height noise was <2 nm peak to peak). At each position, separated by ca. 27 nm, 7938 photocounts were obtained. Error bars on the lifetime data are 1 σ confidence limits on single exponential fits. Far-field power levels: molecule A, ($\bullet$) 22 and ($\blacktriangle$) 2.5 nW; C, ($\bullet$) 38 nW; E, ($\bullet$) 31 and ($\blacktriangle$) 3.8 nW; and G, ($\bullet$) 42 and ($\blacktriangle$) 4.9 nW. Adapted from Ref. 21.

is moved inward away from the metal coating, τ_f increases. The τ_f values do not depend on power (the triangles and circles overlap in Figs. 12(b), (f), and (h) and therefore are not a function of the tip temperature or near-field irradiance. The measured lifetimes vary from below to above the mean bulk lifetime value indicated by the dashed line in Fig. 12. Clearly, the fluorescence lifetime of an R6G molecule on silica is altered as a near-field probe moves over the molecule.

Published simultaneously with the work of Ambrose *et al.* was a paper by Xie and Dunn [24] also reporting on time-resolved near-field single molecule measurements. They studied individual sulforhodamine 101 molecules dispersed on a silica coverslip. In Fig. 13 data are presented showing the lifetime variation versus the lateral displacement of the probe and molecule. The probe–sample separation was approximately 7 nm. The trend is the same as that reported by Ambrose, namely the lifetime is shorter near the metal edges, but the magnitude is somewhat different. In particular, Xie and Dunn observe a lifetime that is everywhere shorter than that measured far-field. When positioned under the center of the aperture the lifetime was found to be 2.0 ns compared to the previously reported value of 2.7 ns, measured on an ensemble, far-field. It was not, and perhaps still is not clear why Ambrose observed lifetimes that were longer than and Xie and Dunn observed lifetimes that were shorter than the unperturbed value when the probe was centered over the molecule. One notable difference in their respective experiments was the probe aperture size, namely about 300 and 125 nm. Another point to consider is that in the case of a molecule near a metal mirror, the lifetime effect is strongly orientation dependent. If the double-lobed structure of the single molecule images in Ref. 21 (reproduced below as Fig. 22) was indicative of a 'z'-oriented molecule, then that would be an

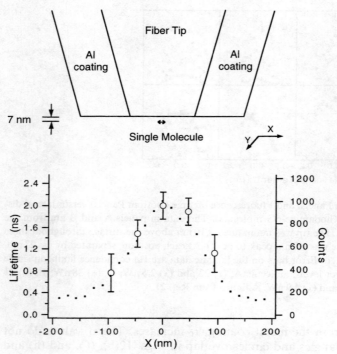

Figure 13. Plot of the lifetime (o) and intensity (■) as a function of x-distance across the emission feature of a single molecule. Above is a schematic of the near-field probe (aperture diameter = 125 nm) with its dimensions drawn on scale with the plot below (curve FWHM = 125 nm). In this example, there is a substantial decrease in the fluorescence lifetime of the molecule near the aperture edges. Adapted from Ref. 24.

important difference between the two experiments. In light of the comment made at the close of Section 2.4.2, however, the orientation of the molecules in Ref. 21 is an open question.

These questions notwithstanding, the situation became even more complicated with the initial results of Trautman and Macklin [18]. When they measured the lifetime of the canonical Bell Labs sample, diI on PMMA, the lifetime was observed to be shorter when the molecule was centered under the aperture relative to that with the molecule near the edge of the aperture (Fig. 14). After repeated measurements, it became clear that this was not due, exclusively, to the difference in the sample, in particular, the presence of the polymer film. For with some probes, the Ambrose/Xie lifetime trend was observed. Macklin and Trautman, consequently, concentrated on probe morphology. Prompted by these results, the Battelle group made measurements as a function of tip–sample separation and found that the lifetime *vs.* lateral displacement curve inverted as the gap was increased from 5 to 20 nm (Fig. 15) [25].

To further the quantitative understanding of the interplay between quenching and spontaneous emission modification and their respective distance dependence, Bian, Dunn, Xie and Leung ran numerical simulations of the experiments [25]. They employed a finite-difference time-domain (FDTD) model on a 2-D lattice, represented in Fig. 16. Results are displayed in Fig. 17 for horizontally and vertically oriented dipoles and for probe–sample gaps of 6 and 24 nm. The computed results show the reversal of lifetime behavior in going from small to large gaps for horizontal dipoles.

Figure 14. (a) Constant-shear-force image of a 100 nm Si sphere, or an image of a probe tip taken with the Si sphere. The Si sphere, being smaller than the probe tip, 'images' the probe tip, rather than the converse. (b) Linecut through the data, as indicated in (a). (c) Fluorescence intensity and lifetime data from a single molecule taken as the molecule was scanned across the tip depicted in (a). In the middle of the aperture, the lifetime is about two times shorter than the unperturbed value. Adapted from Ref. 18.

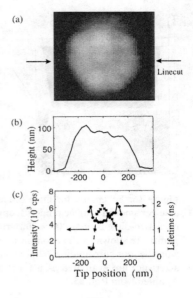

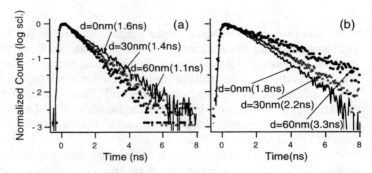

Figure 15. Fluorescence decays of a single molecule measured at three lateral displacements from the tip center $d \approx 0$, 30, and 60 nm. At a tip–molecule gap h of ca. 5 nm, the lifetime is the longest at the tip center and shortened at the tip edge. (b) The opposite trend in lifetime is seen at a larger tip–molecule gap of $h \approx 20$ nm. The lifetime in the farfield without the tip is 2.7 ns. Adapted from Ref. 25.

For horizontal dipoles at small gaps, the quantum yield, q', decreases near the metal edges due to increased nonradiative energy transfer from the molecule to the metal. For horizontal dipoles at larger gaps, the reflected field is out of phase (negative $\phi°$) and the dipole emission is suppressed. For vertical dipoles, the phase of the reflected field is always positive, enhancing the spontaneous emission; consequently, the fluorescence lifetime is always shorter than the lifetime in the absence of the probe. In this case, the decay is dominated not by energy transfer, but by enhanced emission. In all instances, the fluorescence lifetime near the middle of the probe was found to be shorter than the unperturbed value.

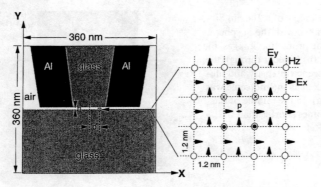

Figure 16. Schematic for finite-difference, time-domain (FDTD) calculations of fluorescent molecule behavior in near-field microscopy. The aperture diameter is 96 nm. The calculation is performed on a two-dimensional 300×300 grid of 1.2 nm square cells. The arrangement of E_x, E_y, and H_z points are shown in the zoom-in inset. A horizontal point dipole is placed at the center of the cell with the four surrounding H_z points driven sinusoidally in the simulation. Molecular emission characteristics are evaluated as a function of the lateral displacement d and the tip–molecule gap h. Adapted from Ref. 25.

As mentioned above, some of the lifetime values for R6G near the center of the probe were longer than the ensemble value measured far-field. This could arise for conditions that are not represented in the calculations, or because there is an inhomogeneous distribution of fluorescence lifetimes for R6G on silica. The latter issue can be addressed in a far-field single-molecule fluorescence lifetime experiment. Such an experiment has recently been reported for diI in PMMA [26], in which orientation and emission-wavelength dependent variations in the unperturbed lifetime were found to be as large as a factor of two.

If near-field is to be used as a measurement tool for intrinsic fluorescence properties, then one needs to know the operating conditions or samples that are appropriate for lifetime and/or spectroscopic measurements. To address this question, the FDTD method was used to compute the spectral shift and fluorescence lifetime as a function of the tip–sample gap with the molecule directly under the center of the aperture. Fig. 18 shows these results. For a γ_0 of about 3×10^8/s, the frequency can red shift by about 10 GHz for distances less than 50 nm. This frequency shift is not important at room temperature where the vibronic bands are approximately 10^4 GHz wide. At low temperatures, the linewidths are in the 0.01 to 10 GHz range, and probe-induced frequency shifts should be measurable. The frequency dependence on distance was not calculated at the larger distances, 200–400 nm, relevant to the experiment of Moerner *et al.* [11] (Section 2.4.2). On the other hand, the fluorescence lifetime of high quantum yield molecules was predicted to be affected in first order throughout the near-field region. Bian *et al.* [25] point out that for systems with fast non-radiative decay channels, the perturbations by the probe may not be significant.

Figure 17. FDTD results for both a horizontal dipole [left column, (a)–(g)] and a vertical dipole [right column, (h)–(n)] as a function of the lateral displacement, d, evaluated at tip–molecule gaps $h = 6$ and 24 nm. From top to bottom, the quantities simulated are the reflected field amplitude E_o and phase ϕ, normalized lifetime τ/τ_o, normalized radiative rate γ_r/γ_o and non-radiative rate γ_{nr}/γ_o, apparent quantum yield q', and spectral shift $\Delta\omega/\gamma_o$. Adapted from Ref. 25.

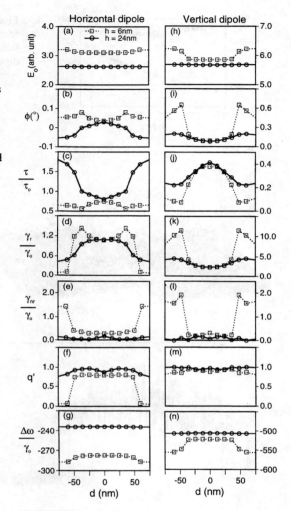

2.4.5 Photobleaching and intensity fluctuations

Because photobleaching limits the total observation time, all reports on room-temperature near-field single molecule detection have noted this disappearing act. Photobleaching is the photoinduced loss of fluorescence, presumably due to a photochemical change of the molecule, such as oxidation. The photobleaching quantum efficiency or branching ratio, ϕ_b, is defined as the probability per optical absorption of generating a non-fluorescent photoproduct. In an ensemble measurement, photobleaching results in a gradual fading of the fluorescence. Note that a non-exponential decrease in the emission intensity, $I(t)$, of the ensemble implies a distribution of ϕ_b's.

In a direct, single-molecule, photobleaching experiment the surface is first imaged

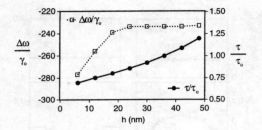

Figure 18. FDTD results for the normalized lifetime τ/τ_o and spectral shift $\Delta\omega/\gamma_o$ as a function of tip-molecule gap h for a horizontal dipole centered on the near-field tip axis. Adapted from Ref. 25.

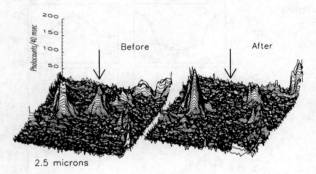

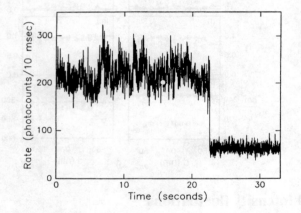

Figure 19. Single molecule photobleaching. Top Left: A $2.5 \times 2.5\,\mu\text{m}^2$ fluorescence image of an R6G-on-silica sample excited in the near-field with 17 nW (as measured in the far-field). The tip was positioned over the molecule labeled by the arrow. Lower Panel: The power was increased by a factor of 10 and the fluorescence intensity recorded. The fluorescence rate was roughly constant, then droped abruptly to the background level. The same region of the sample was re-imaged (Top right), and the molecule under the arrow was found to be missing from the fluorescence image.

to locate the molecules, then the tip is positioned over one molecule and the intensity recorded versus time [27]. Instead of the smooth, monotonically decaying curves observed for high-coverage samples, the signal in Fig. 19 exhibits an approximately constant value, then drops abruptly to the background level. After the signal drops out, the surface is reimaged, and the entire feature at the bleach position is missing. The time between bleaching and reimaging can be long (many tens of minutes), showing that the absence of fluorescence persists. Analogous experiments were performed on single molecules in solids at low temperature [28], but in those cases the scan before and after 'burning' is in frequency rather then position. The reason for the abrupt change is that the molecule is continuously cycled through its ground and excited states by the absorption and re-emission of photons, until during one such cycle it is phototransformed and stops fluorescing. To the chemist, this description is wanting, in that it fails to identify a molecular mechanism behind the photoinduced transformation. A great opportunity exists for the application of single molecule methods in conjunction with chemical techniques to explore the bleaching phenomena. In general, it is not even known whether the molecule stops fluorescing because the photochemically modified species has a larger optical transition energy and is no longer resonant with the excitation radiation or because the photochemistry has produced a species that still absorbs the incident radiation but has a fast radiationless relaxation channel.

Since the near-field tip temperature increases with input power, one possibility for some of the changes observed in bleaching experiments is a sample heating effect. To test for heating effects, 632.8 nm light, which R6G molecules do not absorb, was used to heat the tip during experiments on higher coverage samples. With 3.3 mW of input power at 632.8 nm (180 nW transmitted into the far-field), the tip was set at fixed positions on the sample for as long as 12 min. There were no observable changes in subsequent fluorescence images excited with 514.5 nm light. Hence, the irreversible removal of fluorescence with 514.5 nm light is a photo-induced effect, not a thermal effect.

The quantum efficiency for photobleaching, ϕ_b, may be estimated from the measured single molecule detection rate, R_f, the collection and detection efficiency, D, the fluorescence quantum yield, ϕ_f, and the burn time, $\tau_b/\phi_b = \phi_f/(R_f \times D \times \tau_b)$. Values of ϕ_b range from 10^{-5} to 10^{-8} for molecules on surfaces. In general, such a range is indicative of the heterogeneity of the environments, in particular, the different binding geometries, as has been shown for glass [22].

As discussed above and shown in Fig. 19, some molecules have an emission intensity that is relatively constant prior to photobleaching. Other molecules exhibit more interesting behavior: The fluorescence intensity fluctuates, sometimes abruptly, between different constant levels that last many seconds. On occasion the fluorescence signal can even disappear completely for a number of seconds and then recover. Investigations of intensity fluctuations have been reported on two types of samples: rhodamines on silica, [24, 29] and rhodamines in thin solid polymer films [30]. Fig. 20 shows examples of fluctuating $I(t)$ during direct, stationary observations. Blank traces are shown also for comparison. Intensity fluctuations also can be inferred by extracting the fluorescence intensity that is briefly sampled at widely spaced intervals during successive images (see Fig. 23, below).

Many possible sources can be imagined for the fluctuations including: (1) metastable excited states, such as photoionization with transient electron trapping, (2) metastable complex formation, such as association/dissociation with a nearby molecule, (3) isomerization, including small changes in the nuclear coordinates that result in a spectral shift, (4) changes in the molecular configuration of the environment and a concomitant spectral shift, and (5) reorientation of the optically active molecule.

The latter would result in intensity fluctuations if the excitation were linearly polarized or if the reorientation included an out-of-plane component. In either of these cases, the absorption probability which is proportional to $|E \cdot p|^2$ would change

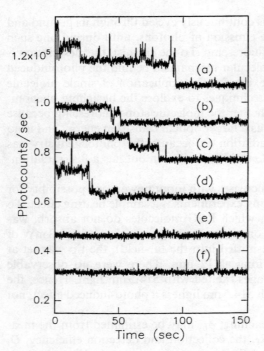

Figure 20. Fluorescence intensity versus time for single R6G molecules on silica. Traces (a)–(d) were taken on single fluorescent spots, and traces (e) and (f) were taken in dark regions (background). The traces are offset by multiples of 1.5×10^4 counts/sec. The photocounts were accumulated in 10 ms bins and later convolved with a 100 ms window. Adapted from Ref. 29.

with transition dipole orientation. Xie and Dunn tested the reorientation hypothesis at the single molecule level for sulforhodamine 101 on silica by alternating the excitation field between two orthogonal linear polarization states [24]. The data are reproduced in Fig. 21. Maximum sensitivity was achieved by aligning one of the polarization axes along p. Whereas the fluctuations in $I(t)$ were large, no fluorescence was observed when the excitation field was orthogonally polarized, ruling out hypothesis (5), at least for this sample. Given that the rate of emission, $R_f = \sigma(I/h\nu)\phi_f$, does change, the fluctuations can be associated with changes in $\sigma(\lambda), \phi_f$, or both. By the time these results were obtained, the spectral shifting behavior had been observed in the emission spectra of DiI on PMMA [17] (see Section 2.4.3). Xie and Dunn suggested that the origin of the fluctuations were due to similar shifts in the absorption spectra and the corresponding changes in $\sigma(\lambda)$, at the laser wavelength. A thorough investigation of this question is presently underway using far-field excitation [19].

When the fluctuations occur on the timescale of several line scans across the sample (≈ 10 sec), the effects of intensity fluctuations and bleaching can be seen directly in the images. An example of a fluorescence image obtained on a low coverage surface is shown in Fig. 22. The two-lobed feature, (A), is a stable fluorescence image formed by a single R6G molecule. By stable it is meant that the fluorescence intensity was reproducible over many minutes. There is a terminated feature at (D) seen as an abrupt loss of signal to the background due to photobleaching, during a line scan. In similar experiments, terminated features were missing in subsequent images.

Figure 21. (A) Emission counts as a function of time (0.2 s bins) collected by centering the near-field probe directly above a single molecule. Sudden jumps were observed before the photobleaching of the molecule. (B) With the tip positioned above another molecule, the excitation polarization was modulated (0.25 Hz) between x- and y- polarizations with a Pockel's cell. The transition dipole of this molecule was oriented along the x-direction, resulting in a large modulation. A constant signal close to the background level was observed for the y-polarized excitation, and there were sudden fluctuations in the emission signal derived from x-polarized excitation. This indicates that molecular reorientation was not the cause of the sudden jumps in emission observed in (A). (C) Excitation spectrum of 200 nM solution of sulforhodamine 101 in methanol. For (A) and (B), the molecules were excited at 594 nm, and the total emission was detected. (Inset) The molecular structure of sulforhodamine 101. Adapted from Ref. 24.

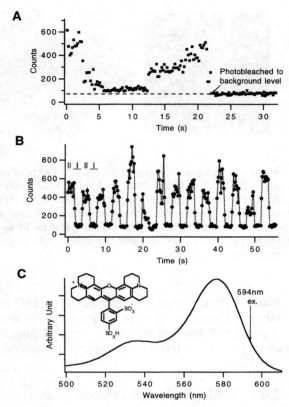

Other molecules showed intensity variations on time scales less than a minute. For example, (B) has a dark line through the lower lobe and (C) has a noisy appearance. A shear force image obtained concurrently with the data in Fig. 22 had no feedback instability other than a 2-nm peak-to-peak height noise throughout the image.

In another set of experiments, A. J. Meixner, *et al.* examined the intensities of single molecules of R6G in a polymer film, rather than on silica [30]. As shown in Fig. 23, the intensities of the spots varied from one image to the next. In more recent experiments [31], they have observed both changes in the shape of the fluorescence image and displacements of its centroid. The conclusion drawn from these experiments is that the R6G molecules in the polymer film, polyvinylbutyral (PVB), are undergoing slow positional diffusion at room temperature, and the corresponding reorientations produce the shape and fluorescence intensity changes.

A third variety of intensity fluctuation has been observed in room-temperature single-molecule experiments, namely that due to real-time observation of single-molecule intersystem crossing (ISC) [32]. The images presented in Fig. 24 were obtained with far-field rather than near-field excitation. In all the other data presented above, whether $I(t)$ plots or images, the bin/pixel time was substantially longer

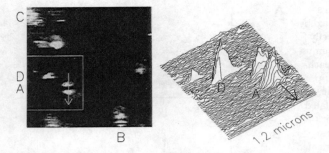

Figure 22. Left panel: Near-field fluorescence image of single R6G molecules on silica. The image size is $2.6 \times 2.7 \, \mu m^2$ (200×200 pixels). Right panel: Surface plot of the box shown left, scanned from left to right, starting at the top. The power in the far field was 34 nW, and the polarization was vertical. The peaks in the surface plot are app. 250 counts per 20 ms high. The single molecule features are ca. 200 to 350 nm wide, indicating that the tip aperture is about 300 nm . The arrows show the path across the two-lobed image (feature A) of a single R6G molecule used for lifetime measurements. Adapted from Ref. 21.

than the triplet lifetime, consequently, the ISC events were not observed. The minimum pixel time is determined by the shot-noise on the fluorescence signal. Because unperturbed data can be taken at much higher flux resulting in higher fluorescence count rates using far-field excitation, the triplet dynamics can be observed.

2.4.6 Biological applications

The final experiment to be presented provides an example from an area of growing interest, that is, the application of near-field spectroscopy at the single molecule level to a sample of biological interest. Over the past few years in a number of experiments, researchers have begun to apply near-field optical methods to biological samples and biomolecules [33–36]. In particular, Ha *et al.* [33] obtained near-field excited fluorescence spectra of single donor–acceptor pairs. Fluorescence resonance energy transfer, or Förster transfer, from a donor to an acceptor is a widely used technique in biophysics for determining the distance between two chromophores, usually dye tags. The physics underlying the technique is the same as that responsible for the quenching of molecular emission near a metal, discussed in Section 2.4.4. In this instance, the acceptor molecule (partially) quenches the emission of the donor when brought into the near-field of the donor. The efficiency of the energy transfer depends not only on the distance, but on the spectra of the two molecules and their relative orientations. As the latter properties will vary from pair to pair, the reason for a single molecule study is clear. The model system used by Ha *et al.* consisted of single tetramethylrhodamine (TMR, the donor) and single Texas Red (TR, the acceptor) molecules covalently linked to the $5'$ ends of hybridized, complementary strands of DNA of lengths 10 or 20 bases. A correlation plot of peak wavelength *vs.*

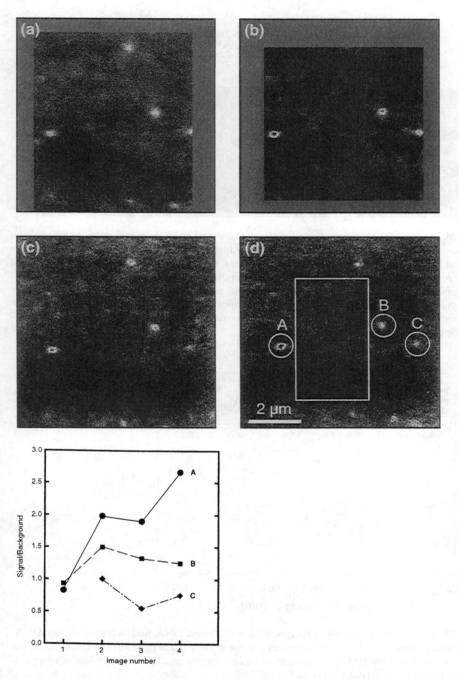

Figure 23. Successive fluorescence excitation images taken from the same area of a thin (ca. 25 nm) polymer film (PVB) doped with a low concentration of R6G molecules. The sample was raster scanned at a tip–sample separation of 5 nm and illuminated with 10 nW (as measured in the far-field) of 514.5 nm light. The FWHM of the single molecule fluorescence spots is 160 nm. The intensities of the spots labeled A, B, and C are shown in the accompanying plot. Adapted from Ref. 30.

(a)   (b)   (c)

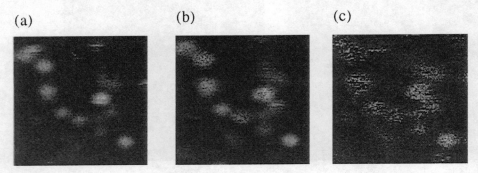

Figure 24. Images of the a group of single molecules obtained under three excitation conditions such that the number of excitation photons per pixel was approximately constant. The excitation power and pixel acquisition times were (a) $0.3\,\mu W$ and $10\,ms$, (b) $4.3\,\mu W$ and $1.0\,ms$, and (c) $33\,\mu W$ and $0.1\,ms$, respectively. At the lowest intensity, the pixel dwell time is longer than the triplet (T_1) lifetime, at the intermediate intensity they are comparable, and at the highest intensity, the T_1 lifetime is longer than the pixel dwell time. The pocked appearance of the molecular images revealed at higher scan rates is due to the molecule having undergone a 'quantum jump' into a dark state, then 'jumping' back to the light state. Adapted from Ref. 32.

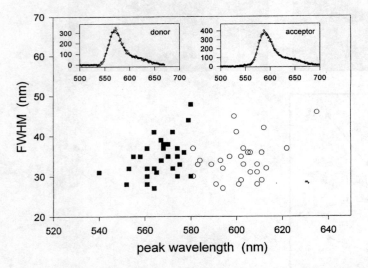

Figure 25. Scatter plot of emission spectra of donor-labeled DNA and acceptor-labeled DNA. Donors and acceptors are indicated by (■) and (○), respectively. Each molecule is represented by the peak wavelength and FWHM of its emission spectrum. Typical spectra, fitted to a sum of two gaussians, are shown as insets. Adapted from Ref. 33.

Figure 26. Top: Background subtracted spectra of three successive scans (5 s/scan) of a single TMR-20-TR pair. The emission is initially acceptor dominated (▲), then mixed (□) and finally appears as donor only (●). Each spectrum is fit with a combination of donor and acceptor spectra, the fits being shown as solid lines. From the fit, it was determined that the acceptor photobleached at the 4th second within the 'middle' scan, and that the energy transfer efficiency was at least 85%. Bottom: Background-subtracted spectra of another TMR-20-TR pair. The first spectrum is plotted with the gray squares, and the second with black circles. In this case, the donor 'died' during the first scan, leaving a reduced acceptor emission in the second and succeeding scans. The solid lines are fits, from which it was determined that the energy transfer efficiency was 53%, assuming equal quantum efficiencies for donor and acceptor. Adapted from Ref. 33.

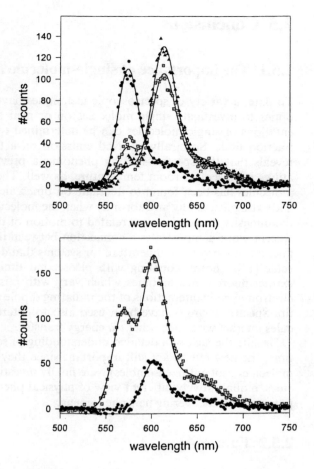

width for single donor and acceptor molecules attached to DNA is presented as in Fig. 25. When the donor and acceptor labeled complementary stands were hybridized then dried onto an amino-derivatized glass surface, spectra such as those displayed in Fig. 26 were observed. The data, which demonstrate that energy transfer was indeed occurring, are discussed in the figure caption. Like many of the other data discussed above, these demonstrate the potential of a technique, rather than answer an outstanding question in an area of research. And as for the potential, a number of issues must be addressed before this method could be generally applied. Chief among these is developing a microscope that can work in a 'biochemical environment'. In addition, the lifetime perturbation due to the near-field probe needs be worked out. Finally and most seriously, to extract an interchromophore distance from the energy transfer efficiency one needs not only the spectra of the donor and acceptor, and their respective orientations in the lab frame (which can be, but was not measured in this experiment), but also the orientation of the intermolecular separation vector (which we don't believe can be measured, presently).

2.5 Conclusions

2.5.1 The importance of single-molecule detection

To date, a variety of spectroscopic techniques have been combined with near-field optics to investigate single molecules on or near surfaces. The locations and orientations of single molecules can be determined with a resolution beyond the diffraction limit. Spectrally-resolved emission on a timescale of seconds to minutes reveals that the spectral-shifting phenomena previously observed at low temperatures is present at room temperature, as well. The fluorescence intensity of single molecules has been found to fluctuate on timescales from microseconds to minutes and extinguish finally and abruptly when the molecule photobleaches. In some cases, the intensity fluctuations are related to motion of the molecules, in other cases not. Recent work demonstrates a connection between the spectral shifting and intensity fluctuations at room temperature for systems that do not involve motion of the molecule [19]. Photon counting with picosecond timing resolution has been used to extract fluorescence lifetimes which vary with position near the probe due to the electromagnetic interactions of the radiating dipole and the dielectric structure of the tip. Spectral changes have been used also to determine when closely spaced molecules interact with each other *via* energy transfer.

Despite the lack of a detailed understanding of some of these observations at this time, the new effects are still important since they were only observed once single molecules (not entire ensembles) were finally investigated. By observing single molecules, a qualitatively different view of physical phenomena is available, and there is potential for uncovering new phenomena.

2.5.2 Future

Near-field scanning optical microscopy was the technique first used to demonstrate that single molecules could be detected and probed at room temperature in an imaging microscopy. Whereas it has been shown recently [26, 32, 16] that far-field is the better choice for studying 'artificial' single-molecule samples such as DiI on PMMA, there are many 'real' samples that require the unique combination of single molecule sensitivity and high spatial resolution afforded by near-field optics. We cite here two examples. The first is the work of Betzig *et al.* [34] wherein fluorescence images were obtained of the actin network of intact fibroblast cells. The resolution and sensitivity of the near-field excited images was demonstrably superior to those of the best confocal images. This area of biomembrane imaging remains one of the frontiers of near-field research. The second is the work of the Barbara group [37] on molecular aggregates which possess a good deal of heterogeneity on a 20–200 nm lengthscale. A number of challenges remain, however, in the further development of near-field microscopy. The most general are that the flux is not yet high enough for rapid imaging, the shear-force feedback is not yet sufficiently flexible for all applications, and the tip perturbation phenomenon needs to be further characterized. This said, the notion that researchers are beginning to be able to observe single-entity transformations directly in a range of diverse and controllable environments is very exciting. We have already found new and unexpected phenomena when observing the members of rather than the whole ensemble. As the range of these studies expands, new insights will abound.

References

[1] For a history of the near-field concept see D. McMullan, *Proc. R. Microsc. Soc.* **25**, 127, (1990). For a review of the early experimental work and theoretical underpinnings of near-field optical microscopy see D. W. Pohl, in *Advances in Optical and Electron Microscopy* (Eds. C. J. R. Sheppard and T. Mulvey) Academic Press, London, 1990, p. 243.

[2] E. A. Ash and G. Nicholls, *Nature* **237**, 510, (1972).

[3] This is covered in many optics texts, including: S. G. Lipson, H. Lipson, and D. S. Tannenbaum, *Optical Physics*, 3rd ed., Cambridge University Press, 1995, Ch. 12; J. W. Goodman, *Introduction to Fourier Optics*, McGraw-Hill, 1968, Ch. 6.

[4] E. Betzig, J. K. Trautman, T. D. Harris, J. S. Weiner, and R. L. Kostelak, *Science* **251**, 1468, (1991). The first near-field optical probes were also embodiments of the aperture idea: D. W. Pohl, W. Denk, and M Lanz, *Appl. Phys. Lett.* **44**, 651, (1984); A. Lewis, M. Isaacson, A. Harootunian, and A. Muray, *Ultramicroscopy* **13**, 227, (1984).

[5] E. Betzig and J. K. Trautman, *Science* **257**, 189, (1992).

[6] H. A. Bethe, *Phys. Rev.* **66**, 163, (1944).

[7] L. Novotny and D. W. Pohl, in *Photons and Local Probes. Prodeedings of the NATO Advanced Research Workshop* (Eds. O. Marti and R Moller), Kluwer, Dordrecht, 1995, p. 21.

[8] G. A. Valaskovic, M. Holton, and G. H. Morrison, *Appl. Opt.* **34**, 1215, (1995).

[9] E. H. Hellen and D. Axelrod, *J. Opt. Soc. Am. B*, **4**, 1337, (1987).

[10] E. Betzig, P. L. Finn, and J. S. Weiner, *Appl. Phys. Lett.* **60**, 2484 (1992); R. Toledo-Crow, P. C. Yang, Y. Chen, and M. Vaez-Iravani, *Appl. Phys. Lett.* **60**, 2957, (1992).

[11] W. E. Moerner, T. Plakhotnik, T. Irngartinger, U. P. Wild, D. W. Pohl, and B. Hecht, *Phys. Rev. Lett.* **73**, 2764, (1994).

[12] K. Karrai and R. D. Grober, *Appl. Phys. Lett.* **66**, 1842, (1995); J. W. P. Hsu, M. Lee, and B. S. Deaver, *Rev. Sci. Inst.* **66**, 3177, (1995).

[13] E. Betzig and R. J. Chichester, *Science* **262**, 1422, (1993).

[14] C. J. Bouwkamp, *Philips Res. Rep.* **5**, 321, (1950).

[15] D. A. Christensen, *Ultramicroscopy* **57**, 189, (1995); L. Novotny, D. W. Pohl, and B. Hecht, *Opt. Lett.* **20**, 970, (1995).

[16] J. J. Macklin and J. K. Trautman, manuscript in preparation.

[17] J. K. Trautman, J. J. Macklin, L. E. Brus, and E. Betzig, *Nature* **369**, 40, (1994).

[18] J. K. Trautman and J. J. Macklin, *Chem. Phys.* **205**, 221, (1996).

[19] X. S. Xie, personal communication.

[20] K. H. Drexhage, *Prog. in Opt.* **12**, 163, (1974); H. Kuhn, *J. Chem. Phys.* **53**, 101, (1970); R. R. Chance, A. Prock and R. Silbey, *Adv. Chem. Phys.* **37**, 1, (1978).

[21] W. P. Ambrose, P. M. Goodwin, J. C. Martin, and R. A. Keller, *Science* **265**, 364, (1994).

[22] A. L. Huston and C. T. Reimann, *Chem. Phys.* **149**, 401, (1991).

[23] J. N. Demas and B. A. DeGraff, *Sensors and Actuators* **B 11**, 35, (1993).

[24] X. S. Xie and R. C. Dunn, *Science* **265**, 361, (1994).

[25] R. X. Bian, R. C. Dunn, X. S. Xie, and P. T. Leung, *Phys. Rev. Lett.* **75**, 4772, (1995).

[26] J. J. Macklin, J. K. Trautman, T. D. Harris, and L. E. Brus, *Science* **272**, 255, (1996).

[27] W. P. Ambrose, P. M. Goodwin, J. C. Martin, and R. A. Keller, *Phys. Rev. Lett.* **72**, 160, (1994).

[28] Th. Basché and W. E. Moerner, *Nature* **355**, 335, (1992).

[29] W. P. Ambrose, R. L. Affleck, P. M. Goodwin, R. A. Keller, J. C. Martin, J. T. Petty, J. A. Schecker, and M. Wu, *Experimental Technique of Physics* in press, (1996).

[30] A. J. Meixner, D. Zeisel, M. A. Bopp, and G. Tarrach, *Opt. Eng.* **34**, 2324, (1995).

[31] A. J. Meixner, personal communication.

[32] J. K. Trautman, *Proceedings of the Robert A. Welch Foundation 39th Conference on Chemical Research: Nanophase Chemistry.* Welch Foundation Press, Houston, 1996.

[33] T. Ha, Th. Enderle, D. F. Ogletree, D. S. Chemla, P. R. Selvin, and S. Weiss, *Proc. Natl. Acad. Sci. USA* in press, (1996).

[34] E. Betzig, R. J. Chichester, F. Lanni, and D. L. Taylor, *Bioimaging* **1**, 129, (1993).

[35] R. C. Dunn, G. R. Holtom, L. Mets, and X. S. Xie, *J. Phys. Chem.* **98**, 3094, (1994).

[36] R. C. Dunn, E. V. Allen, S. A. Joyce, G. A. Anderson, and X. S. Xie, *Ultramicroscopy* **57**, 113, (1995).

[37] D. A. Higgins, P. J. Reid, and P. F. Barbara, *J. Phys. Chem.* **100**, 1174, (1996); D. A. Higgins, J. Kerimo, D. A. Vanden Bout, and P. F. Barbara, *J. Am. Chem. Soc.* in press, (1996).

3 Single-Molecule Detection in Analytical Chemistry

N. J. Dovichi, D. D. Chen

3.1 Introduction

Chemical analysis conventionally entails two tasks. The first, qualitative analysis, deals with the identification of an unknown substance, typically in a complex environment. Inevitably, some sort of separation technique, be it chromatographic, electrophoretic, or immunological, is employed when analyzing biological substances. The second, quantitative analysis, deals with the determination of the amount of analyte present. Inevitably, some sort of spectroscopic, electrochemical, or mass spectroscopic method is employed to determine the amount of substance present in the sample following the qualitative analysis step.

Classically, quantitative analysis is an estimation process, where the amplitude of some signal is used to estimate the concentration or amount of analyte within a sample. Single molecule detection represents a fundamentally different approach to quantitative analysis, where the presence of an analyte molecule is detected and counted. This binary signal processing inherently provides the highest level of accuracy – a sample can be analyzed with no higher accuracy than by counting every analyte molecule within the sample. The precision of replicate analyses will be limited by shot noise in the average number of molecules contained within the sample.

This chapter considers the quantitative aspects of single-molecule detection in solution by laser-induced fluorescence. We will describe the combination of single molecule detection with capillary electrophoresis as the ultimate analytical procedure. Other methods of single molecule detection, such as electrochemical detection, are not considered [1, 2].

3.1.1 Photon-burst detection of single molecules

In laser induced fluorescence detection of single molecules, a dilute suspension of fluorescent analyte is passed through a focused laser beam, Fig. 1. As the molecule passes through the beam, a burst of fluorescence is observed. This burst is generated as the molecule absorbs and emits many photons during transit through the beam. The duration of the single molecule fluorescence bursts is given by the residence time of the molecule in the laser beam. This residence time can be dominated by the transit time of the molecule through the laser beam, the diffusion time of the mole-

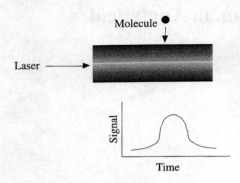

Figure 1. Single-molecule detection. A molecule flows through the focused laser beam, generating a photon burst, which is shown as the time dependent signal at the bottom of the figure. The burst must be detected above the noise in the background signal in a single molecule detection experiment.

cule through the beam, or the photobleaching time of the molecule. The total number of photons contained in the photon burst is related to the spectral properties of the molecule, the residence time of the molecule in the laser beam, and the laser irradiance. Laser irradiance cannot be increased without limit; photobleaching limits the total number of photons emitted per molecule and saturation limits the maximum emission rate.

The photon burst must be detected above the noise in the background signal. The background signal may be due to Raman and Rayleigh scatter from the solvent. fluorescence from impurity molecules, scatter and fluorescence from the windows, and detector dark current. The noise in the background signal is usually dominated by shot noise in the number of background photons.

Typically, the signal is filtered and treated with a threshold algorithm. Those peaks that are larger than the threshold are counted as a molecule, while those peaks that fall below the threshold are ignored. There is a trade-off in setting the threshold. If the threshold is set too low, noise spikes from the background can exceed the threshold and generate a false alarm. The false alarm rate is the background count rate generated by the reagent blank; this false alarm rate may be due to either solvent contamination or noise in the background signal. Pulsed laser excitation and time gated detection have been used to discriminate light scatter from fluorescence from an analyte molecule, reducing the false alarm rate. The use of dyes that absorb in the red or near-infrared results in a lower false alarm rate compared with blue emitting dyes, simply because there tend to be fewer contaminants that can absorb in this spectral region.

Another type of error occurs when the signal from a molecule does not exceed the threshold, generating a miss. The detection efficiency is the ratio of detected molecules to total analyte molecules in the sample. Detection efficiency is degraded in three ways. First, if the detection threshold is set too high, then only a small fraction of the molecules will generate detectable photon bursts. Second, at high concentrations, several molecules may be present in the probe volume at the same time. The simple threshold counter is not able to resolve multiple molecules and the detection efficiency drops compared to low analyte concentration data. Third, if only a small portion of the sample passes through the detection volume, then most molecules can not be detected. For example, if a 10 μm diameter laser beam is used to illuminate a 1 mm diameter flow cell, then only a very small fraction of analyte molecules will pass through the laser beam.

Low detection efficiency leads to ambiguous analytical results. If the poor detection efficiency is caused by a mismatch between the beam spot-size and the flow chamber dimensions, then molecules that pass through the edge of the probe volume will generate small photon bursts, which are difficult to distinguish from background noise. Two strategies have been used to ensure that all molecules pass through the probe volume. A sheath flow cuvette has been used wherein a thin sample stream is surrounded by flowing buffer; the sheath fluid forces all of the analyte to pass through the laser beam. Alternatively, the analyte stream has been broken into individual droplets; these droplets are then analyzed individually.

3.1.2 Digital communication theory

Digital communication engineers have faced the problem of setting the appropriate threshold level in fiber-optic communication application where the presence of a pulse of light must be discriminated against background noise [3, 4]. The signal processor must discriminate between two hypotheses: the null hypothesis, H_0, that no molecule is present in the probe volume, and the alternative hypothesis, H_1, that one (or more) molecule(s) is present. Continuing with the communication nomenclature, a false alarm occurs when H_1 is declared when H_0 is true. A miss occurs when H_0 is declared when H_1 is true. With a particular threshold, there will be some probability of a false alarm, P_{fa}, and some probability of a miss, P_m. The threshold must be chosen to minimize these error probabilities.

During an observation period, H_0 will be true with some probability P_0 and H_1 will be true with probability P_1, and $1 = P_0 + P_1$. In single-molecule experiments, P_1 corresponds to the fraction of time that a molecule is present in the probe volume; in other words, P_1 is the fractional occupancy of the probe volume. In general, this probability is unknown at the start of the experiment and is often the quantity sought.

During the experiment, the signal is sampled at one or more points, creating a data vector N, which consists of the number of photocounts observed during each point in the acquisition window. Based on the vector, a decision must be made between H_0 and H_1. The data points are conditionally distributed according to some law. If H_0 is true, then the conditional probability is $P(N|H_0)$. If H_1 is true, the conditional probability is given by $P(N|H_1)$.

Due to a lack of space, we describe a very simple data treatment. The photocounts contained in the sample vector are summed to yield one datum, N. We also assume that the signals are Poisson-distributed, so that the probability of observing N photocounts given that no molecule is present is given by

$$P(N|H_0) = \frac{\mu_b^N e^{-\mu_b}}{N!} \tag{1}$$

and the probability of observing N photocounts given that a molecule is present in the probe volume is given by

$$P(N|H_1) = \frac{(\mu_b + \mu_s)^N e^{-\mu_b + \mu_s}}{N!} \tag{2}$$

where μ_b is the mean background photocount and μ_s is the mean signal photocount. The average number of photocounts when a molecule is present is given by the sum of the background photocounts plus the number of photocounts from the molecule. The ratio of the conditional probabilities is called the likelihood ratio

$$\Omega = \frac{P(N|H_1)}{P(N|H_0)} = \left(\frac{\mu_b + \mu_s}{\mu_b} \right)^N e^{-\mu_s} \tag{3}$$

There are several approaches to setting the threshold. In the Bayes approach, a threshold is chosen to minimize the total probability of an error, given by $P_e = P_0 \times P_{fa} + P_1 \times P_m$. In general, P_0 and P_1 are unknown at the start of the experiment, but may be estimated from preliminary experiments. P_e is minimized by the following detection strategy. The processor will declare H_1 is true if

$$\Omega = \frac{P(N|H_1)}{P(N|H_0)} = \left(\frac{\mu_b + \mu_s}{\mu_b} \right)^N e^{-\mu_s} > \frac{P_0}{P_1} \tag{4}$$

and the processor will declare H_0 true if

$$\Omega = \frac{P(N|H_1)}{P(N|H_0)} = \left(\frac{\mu_b + \mu_s}{\mu_b} \right)^N e^{-\mu_s} < \frac{P_0}{P_1} \tag{5}$$

Since the logarithm is monotonic for positive numbers, the probabilities can be replaced by their logarithms, and H_1 is declared if

$$-\mu_s + N \ln \left(\frac{\mu_b + \mu_s}{\mu_b} \right) > \ln \left(\frac{P_0}{P_1} \right) \tag{6}$$

Solving for N, we get that H_1 is declared if

$$N > \frac{\ln \left(\dfrac{P_0}{P_1} \right) + \mu_s}{\ln \left(\dfrac{\mu_b + \mu_s}{\mu_b} \right)} = L \tag{7}$$

where L is the decision threshold. That is, the detection scheme consists of comparing the number of photocounts with a threshold L, and declaring a molecule present if the photocounts exceed the threshold. Note that the decision threshold depends on the logarithm of the ratio of P_0/P_1. As a result, a rough value of P_1 may be estimated and used to set the threshold without much decrease in precision.

The probability of a false alarm is given by the probability that the number of photocounts from the background exceeds the threshold

$$P_{fa} = \sum_{L}^{\infty} \frac{\mu_b^N e^{-\mu_b}}{N!} \tag{8}$$

The probability of a miss is given by the probability that the number of photons from the molecule fall below the threshold

$$P_m = \sum_{0}^{L} \frac{(\mu_b + \mu_s)^N e^{-\mu_b + \mu_s}}{N!} \tag{9}$$

This model assumes that the amplitude from a single-molecule burst is a simple Poisson distributed function. However, in most single-molecule experiments, the amplitude is doubly-stochastic in that the expected number of photons in the burst will depend on the path of the molecule through the laser beam; those molecules that pass through the center of the beam will generate larger signals, on average, than the molecules that pass through the wings of the beam [5]. Nevertheless, the simple Bayesian model provides a convenient starting point in the discussion of the performance of the detection systems. Finally, at high concentrations, the simple single molecule detector will saturate because it cannot distinguish between one and many molecules. The detector is said to be paralyzable because it cannot respond to the next molecule until the first has passed the probe volume. Analysis of the width of the fluorescence burst can be used to partially correct for two molecules passing through the detector in quick succession [6]. It is possible to use a set of discrimination levels to estimate the number of molecules present in the interaction region [7].

3.3.1 Sample throughput

Single-molecule detection experiments are characterized by one other parameter, sample throughput, which is the sample volume analyzed per second; in general, a decrease in sample throughput results in improved signal-to-noise ratio for single molecule detection [8]. On the other hand, a high sample throughput is necessary to analyze very dilute solutions [9]. For example, there is, on average, one molecule per nanoliter in a $1.6 \times 10^{-15}\,\mathrm{mol\,l^{-1}}$ solution; many nanoliters must be analyzed to count enough molecules to estimate precisely the analyte concentration. If the sample throughput is in the femtoliter per second range, an unacceptably long period is required to analyze a sufficient sample volume to characterize the dilute sample.

3.2 History of high-sensitivity fluorescence detection in flowing streams

In general, signal-to-noise improves for single molecule detection with a decrease in the probe volume. Heuristically, single molecule detection is a binary decision process: either a molecule is present or absent from the probe region. When a molecule is absent from the probe volume, the analyte concentration is effectively zero in the probe region. When a molecule is present in the probe volume, the effective analyte concentration is inversely proportional to the probe volume. One molecule in a very small probe volume results in a very large effective analyte concentration, which is relatively easy to discriminate against the zero-concentration background signal. Unfortunately, a small probe volume usually results in a very low sample throughput, which presents difficulties in analyzing real samples.

3.2.1 Total internal reflection microscopy

In 1976, T. Hirschfeld used total internal reflection excitation in a confocal fluorescence microscope to achieve a 3 fL probe volume; this very small probe volume allowed detection of individual antibodies, labeled with 100 fluorescent molecules. The sample was spread as a thin film of sample on a microscope slide [10]. As the microscope slide was translated across the laser beam, bursts of fluorescence from single molecules were detected above the noise in the background signal. The signal-to-noise ratio of this experiment was very high; assuming Poisson distributed background photocounts, the signal-to-noise ratio exceeded 40 for single antibody detection. However, this high signal-to-noise ratio was achieved only for molecules passing through the center of the laser beam. Molecules that passed off-axis through the beam would generate smaller peaks with correspondingly low signal-to-noise ratio. The detection efficiency is low in the total-internal fluorescence experiment because a small fraction of the analyte molecules are found in the thin region probed by the evanescent wave created by the total-internal reflection excitation beam.

In Hirschfeld's data, the background signal was about 100 photocounts, while the photon burst amplitude was about 400 photocounts larger. That is, $\mu_b = 100$ and $\mu_s = 400$. Also, the probability that a molecule was present in the laser beam looks to be about $P_0 = 0.2$; by definition, the probability that a molecule was not present is $P_0 = 0.8$. Using the equations from above, the optimum threshold, in a Bayes sense, is given by

$$N > \frac{\ln\left(\dfrac{P_0}{P_1}\right) + \mu_s}{\ln\left(\dfrac{\mu_b + \mu_s}{\mu_b}\right)} = \frac{\ln\left(\dfrac{0.8}{0.2}\right) + 400}{\ln\left(\dfrac{100 + 400}{100}\right)} = 250 = L \qquad (10)$$

The false alarm rate is given by

$$P_{fa} = \sum_{250}^{\infty} \frac{100^N e^{-100}}{N!} = 4 \times 10^{-36} \tag{11}$$

while the miss rate is given by

$$P_m = \sum_{0}^{249} \frac{(500)^N e^{-500}}{N!} = 1 \times 10^{-35} \tag{12}$$

These very small error rates could never be achieved in practice. First, as noted above, molecules passing off-axis will generate low amplitude peaks with lower detection probability; the simple model grossly underestimates the miss rate. More importantly, background contamination will be many orders of magnitude higher than the false alarm rate. This false alarm rate predicts that only one noise spike in about 10^{35} probe volumes will exceed the threshold, or one error for every 10^{21} liters (1000 cubic kilometers) of sample processed. Inevitably, the real-world problem of sample contamination will dominate the false-alarm rate, rather than noise in the background signal. It has proven to be very difficult to achieve contamination levels less than about 10^{-17} mol l^{-1}, or 6 million contaminant molecules per liter of solution.

3.2.2 Confocal microscopy in solution

Confocal microscopy has been applied to the detection of single molecules in solution. In 1994, Rigler and Mets reported the detection of fluorescence bursts from diffusing single Rhodamine-6G molecules [11]. An extraordinarily small probe volume was produced by the confocal instrument, 0.24 fL. Photon bursts were detected as single dye molecules diffused through the detection volume of the microscope. The extremely small probe volume produced by the microscope is advantageous in achieving good signal-to-noise ratio in single-molecule detection. When a single molecule is present in the probe volume, the effective analyte concentration is 7×10^{-9} mol l^{-1}, which is a relatively high effective dye concentration. However, because of the short illumination time, each molecule generated an average of only 4 detected photons. The background signal under the same conditions was 0.004 photons. The noise in the background signal should be about 0.06 photons, which results in asignal-to-noise ratio of 63 for single molecule detection.

The detection threshold for this experiment will depend on P_1, the probability that a molecule is present in the probe volume. Taking an analyte concentration of 10^{-11} mol l^{-1}, the probability that a molecule is found in the probe volume is 0.0014. The threshold for optimum single molecule detection is

$$N > \frac{\ln\left(\dfrac{P_0}{P_1}\right) + \mu_s}{\ln\left(\dfrac{\mu_b + \mu_s}{\mu_b}\right)} = \frac{\ln\left(\dfrac{0.9986}{0.0014}\right) + 4}{\ln\left(\dfrac{4.004}{0.004}\right)} = 1.5 = L \tag{13}$$

The probability of a false alarm is given by

$$P_{fa} = \sum_{2}^{\infty} \frac{0.004^N e^{-0.004}}{N!} = 8 \times 10^{-6} \tag{14}$$

while the probability of a miss is given by

$$P_m = \sum_{0}^{1} \frac{(4)^N e^4}{N!} = 0.01 \tag{15}$$

While the signal-to-noise ratio is similar for this experiment and that of Hirschfeld, the probability of both a false alarm and a miss is much higher in the confocal microscopy case. The difference reflects the behavior of the Poisson distributed data for small numbers of counts. With a small mean number of photons emitted by a molecule, there is a relatively high probability that the molecule will not emit enough photons to exceed the threshold.

The signal-to-noise degraded as the probe volume increased. A 60 fL probe volume generated a fluorescence signal of 100 photoelectrons due to the longer detection period during which the diffusing molecule remained in the probe region. The signal-to-noise ratio was 20 in this case. which implies a background count of 25 photons. Very similar results were reported by Zare's group [12].

Again, we can compute thresholds and error rates. Again assuming an analyte concentration of 10^{-11} mol l^{-1}, P_1 becomes 0.36. The threshold is then

$$N > \frac{\ln\left(\frac{P_0}{P_1}\right) + \mu_s}{\ln\left(\frac{\mu_b + \mu_s}{\mu_b}\right)} = \frac{\ln\left(\frac{0.64}{0.36}\right) + 100}{\ln\left(\frac{125}{25}\right)} = 62 = L \tag{16}$$

and the false alarm probability is

$$P_{fa} = \sum_{62}^{\infty} \frac{25^N e^{-25}}{N!} = 3 \times 10^{-10} \tag{17}$$

while the probability of a miss is

$$P_m = \sum_{0}^{61} \frac{(125)^N e^{-125}}{N!} = 1.6 \times 10^{-10} \tag{18}$$

It is interesting that the error rates are much lower for the large sample volume. In this case, the error rates drop because the fluorescence signal is sufficiently large to usually exceed the detection threshold. Comparison of these two experiments leads to the interesting observation that the error rate is much lower for the experimental condition that has poorer signal-to-noise ratio.

3.2.3 Levitated droplets

Ramsey at Oak Ridge has demonstrated detection of molecules suspended in levitated microdroplets. In these experiments, droplets of dilute solution were electrostatically levitated at the focus of a laser beam [13–16]. Fluorescence from the droplet was observed to decay with time due to photobleaching. Comparison of the bleached and unbleached signal was used to estimate the number of molecules in the droplet. Detection efficiency should be near unity; every droplet was interrogated for a sufficient time period to ensure detection. Sample throughput was very low. Each 35 fL droplet required about 10 minutes to capture and analyze, corresponding to a sample throughput of 0.06 fL/s. More recent work presented detection of single rhodamine 6G molecules suspended in glycerol droplets with signal to noise ratio of >40 [15].

This extraordinary small probe volume allows high efficiency detection of individual molecules but at the expense of extraordinarily low throughput. About 200 000 years would be required to analyze the content of a 1-μL sample with the early-generation instrument.

3.2.4 Modified flow cytometry – hydrodynamic focusing

Keller's group at Los Alamos introduced the sheath-flow cuvette for high sensitivity detection in flowing neat solutions [17–21]. In these experiments, hydrodynamically focused flow constrains the analyte molecules to pass through the center of the flow chamber. Because the entire sample stream passes through the probe volume, detection efficiency should be quite high. Furthermore, because the sample stream is located far from the cuvette walls, background light scatter is minimized, improving signal-to-noise ratio and reducing the false-alarm rate. Finally, relatively large sample throughput has been achieved with this technique.

Early experiments demonstrated zeptomole detection limits for rhodamine 6G [17, 18]. Later, a dilute solution of B-phycoerythrin was detected as it passed through a laser-beam at a rate of 4 000 molecules/second [19]. Since data were processed in 180-μs bin, on average 0.7 molecules were present in the probe volume. In another experiment, rhodamine 6G was detected at a concentration of 10^{-15} mol l^{-1} with pulsed-laser excitation and time-gated detection of fluorescence; no photon bursts from single molecules were detected [20]. As the first unambiguous single molecule detection in a sheath flow cuvette, DNA restriction digests were labeled with the intercalating dye TOTO [21]. Fluorescence pulses from fragments of 10 086, 17 053, 29 946, 38 416, and 48 502 base pairs were generated with a 30 mW argon ion laser beam. The amplitude of these pulses was used to estimate the fragment size. The labeling dye was incorporated at the rate of about one dye molecule per five base pairs. As a result, each DNA molecule contained several thousand dye molecules. This system had a sample throughput of 6 nL/s. The detection efficiency approached 100%; each analyte molecule passed through the laser beam, generating a detectable pulse well above the background signal. Finally, the false alarm rate appears to be on the order of 0.2 molecules/s for fragments of 10 000 bases in size. Similar work has been published by a second group at Los Alamos [22] and by a group in Berkeley [23].

3.2.5 Large flow chambers

In a second series of experiments from Keller's group, single molecules were detected in bulk flow through a cuvette; sheath-flow was not used in these experiments, which resulted in poor detection efficiency. In the first experiment, a mode-locked argon ion laser was used to excite individual molecules of rhodamine 6G [24]. Sample throughput was about 44 pL/s through the probe volume. Sample transit time through the probe volume was 10 ms, which lead to a maximum count rate of about 100 molecules/s. Detection efficiency was very small; roughly 0.0002% of the sample in the $10 \times 4 \, mm^2$ flow chamber passed through the probe volume. The false alarm rate was $0.01 \, s^{-1}$. In a second experiment, a 500 mW CW laser was used to excite fluorescence from rhodamine 6G in ethanolic solutions [25]. The photostability of rhodamine is enhanced in ethanol, leading to improved signal-to-noise ratio compared with aqueous solutions. Autocorrelation analysis produced evidence that single molecules passed through the laser beam. Unfortunately, discrete photon bursts from individual molecules were not detected. In recent experiments, the pulsed-laser system was used to measure the fluorescence lifetime of individual analyte molecules, including rhodamine 110, fluorescein, and BODIPY [26]. Sample transit time through the laser beam was ca. 4 ms, corresponding to a sample throughput of a few hundred molecules per second. A 700 µm, square-bore flow cell was used, generating a detection efficiency of 0.02%. The false-alarm rate appeared very low, but was not specified.

Mathies reported analysis of $2 \times 10^{-13} \, mol \, l^{-1}$ B-phycoerythrin in a 0.5 mm × 0.5 mm flow chamber [27]. An autocorrelation signal was generated by averaging the signal over a 56 second period. This autocorrelation signal generated a peak at short delays, corresponding to the presence of molecules passing through the beam. The pulse-height distribution showed a slight tail at higher counts, which was interpreted as being caused by single molecules. Analysis of the data revealed that a single B-phycoerythrin molecule generated ca. 12 photons compared with 8 photons generated from a background trace. The sample throughput through the probe volume was about 150 pL/s.

Soper has reported single-molecule detection of an infrared fluorescent dye molecule with excitation by a mode-locked, titanium–sapphire laser [28, 29]. A

0.8 × 0.8 mm flow cell was used. Less than 0.01% of the molecules passed through the 0.8 pL probe volume. The volumetric flow rate through the probe region was 80 pL/s. The false positive rate was 0.02 per second. Similar experiments were reported by Winefordner's group for detection of IR140 analyte flowing in a thin capillary [30].

3.3 Applications of single-molecule detection – DNA sequencing

Keller's group has proposed a very clever method for DNA sequencing that is based on single molecule detection [31, 32]. In their method, a single strand of DNA is replicated in the presence of fluorescently labeled deoxynucleotides. The DNA polymerase makes a faithful copy of the original DNA but with fluorescently labeled nucleotides. A single molecule of this DNA strand is suspended in a sheath flow cuvette. A DNA exonuclease is added to the DNA. This enzyme attaches to the end of the strand of DNA and slowly crawls up the strand, cleaving successive nucleotides, which flow down stream. Each fluorescent nucleotide is transported by the stream through the probe volume of a single molecule detector. As the molecule passes through the beam, a burst of fluorescence is generated and characterized.

In Keller's proposal, four different fluorescent labels are used to tag the four nucleotides that constitute DNA. As each nucleotide is successively cleaved from the strand of DNA, the nucleotide falls through the fluorescence detector, where the fluorescent label is identified based on its fluorescent spectrum and lifetime. Since each nucleotide has a unique fluorescent tag, it is possible to reconstruct the original DNA sequence based on the spectral characteristics of the train of single molecules passing through the probe volume.

This method of DNA sequencing presents several potential advantages over conventional sequencing technology. Once the exonuclease begins to digest the DNA strand, it tends to remove several thousand nucleotides before falling from the strand. As a result, single molecule sequencing will generate long stretches of sequence, which can be very difficult to obtain by conventional means. The technique should be fast since the nuclease can cleave many nucleotides per second.

Single molecule DNA sequencing is not yet a reality, but much progress has been made on the technology. Advances have appeared in the synthesis of the fluorescent DNA fragments, in the detection and discrimination of different fluorophores at the single molecule level, in the introduction of a single strand of DNA in a flowing stream, and in the detection of nucleotides cleaved from the strand. The current bottleneck appears to be the incorporation of four different dyes into the nucleotides and the discrimination of the four dyes in the instrument. However, steady progress in occurring in the project, and single molecule sequencing may become the method of choice in the 21st century.

3.4 Applications of single-molecule detection – capillary electrophoresis

Capillary electrophoresis is a very powerful analytical technique for the separation of complex biological mixtures into individual components [33, 34]. In the experiment, a small slug of analyte is introduced into one end of a buffer-filled capillary. The capillaries are typically 10 to 100 μm inner diameter, about 30 cm in length, and constructed from fused silica. A high potential is applied across the capillary, which drives the analyte through the buffer at a rate that depends on the size and charge of the analyte molecules. The separation speed and resolution increase with applied potential, which typically ranges from 10 to 30 kV. The use of very high potential is facilitated by the narrow dimension of the capillary, which efficiently radiates thermal energy.

My research group has focused on the use of very high sensitivity fluorescence detection in capillary electrophoresis for the analysis of amino acids [35–47], for DNA sequencing [37, 38], for analysis of proteins [39, 40], for analysis of enzymes [41, 42], and for analysis of sugars [43, 44]. These systems provide detection limits from a few hundred molecules down to single molecules of fluorescent analyte. Quite complex mixtures can be separated and analyzed; the present state-of-the-art appears to be resolution of over 600 DNA sequencing fragments in a two hour separation [38].

In this section, we provide some details on the detection of individual molecules of B-phycoerythrin, separated by capillary electrophoresis. B-phycoerythrin is a large protein found in the light harvesting apparatus of certain algae. The protein exists in several different forms, which reflect different decorations added by nature to the peptide backbone of the molecule. However, before discussing the capillary electrophoresis results, we describe the performance of the detector.

3.4.1 Characterization of the instrument

Our single-molecule detector is based on the sheath-flow technology introduced by Keller's group. A 10 μm inner diameter, 142 μm outer diameter, and 29 cm long fused silica capillary is introduced into the flow chamber of a sheath flow cuvette [40]. The volume contained within this capillary is about 25 nL. The locally constructed sheath flow cuvette has 1-mm thick quartz windows and a 150 μm square flow chamber. The flow chamber is held within a stainless steel holder that is held at ground potential. Buffer is introduced into the cuvette, drawing analyte from the tip of the capillary as a fine stream in the center of the flow chamber.

We relied on electrokinetic pumping to introduce analyte molecules into the cuvette at very low rates. The injection tip of the capillary was immersed in a small reservoir that contained B-phycoerythrin at a concentration of 1.7×10^{-12} mol l^{-1}. A platinum electrode was also immersed in the sample, and a several kilovolt potential was applied to the electrode. It is important to encase the electrode in an interlock equipped enclosure to prevent accidental contact with the high voltage.

The electric field drives the B-phycoerythrin through the capillary into the sheath flow cuvette. The analyte molecules drifted from the tip of the capillary and were entrained in the sheath stream. By use of low concentration analyte, narrow diameter capillaries, and a well regulated power supply, the analyte flow rate could be controlled with high precision at very low values. For example, it takes about 125 seconds for a B-phycoerythrin molecule to sweep through the capillary at a driving potential of 29 000 V; the volumetric flow rate at 29 000 V is about 200 pL/s and it scales linearly with potential. We routinely use a driving potential of 1 000 V, which generates a flow rate of 6 pL/s.

Fluorescence is excited by a 2-mW helium–neon laser beam operating in the green at 543.5 nm. The beam is focused to a ca. 30 μm spot about 30 μm below the tip of the capillary. Fluorescence is collected at right angles by two 60 × 0.7 *NA* microscope objectives, located on opposite sides of the cuvette. Fluorescence is imaged onto an iris, passed through a band-pass interference filter and detected with an R1477 photomultiplier tube, operating in analog mode. The current from the two photomultiplier tubes is summed, passed through a 1 kHz low-pass electronic filter and digitized at 1 kHz.

3.4.2 Demonstration of single-molecule detection with a He–Ne

Fig. 2(A) presents the analog photomultiplier signal generated by passing a dilute solution of B-phycoerythrin through the helium–neon laser beam. In this experiment, an electric field was applied to the capillary for the first 30 seconds of data collection; the polarity was then reversed to generate a blank signal; the last ten seconds of data in the figure corresponds to the background signal. The background signal appears to be due to scatter from several weak water Raman bands that overlap with the bandpass of our interference filter. The noise in the background appears to be Poisson distributed, corresponding to a detected background photon flux of about 26 kHz.

Fig. 2(B) presents the probability distribution for the background (dashed curve) and for the sample flowing (smooth curve) at a rate of 6 molecules per second. The width of the background distribution arises from shot noise in the background light scatter signal. The distribution generated by the flowing sample consists of higher amplitude signals arising from the analyte molecules resting upon the background noise distribution.

The distribution of signal from the molecules is quite broad, which reflects several processes. First, shot noise in the number of photons leads to non-uniform peak height. Second, a distribution in peak height arises as molecules pass through different portions of the Gaussian laser beam profile. Third, the analyte used in this case is a protein that has been isolated from bacteria. The protein exists in several different forms, reflecting variations in the glycosylation pattern of the molecule; these different glycoforms may have different spectroscopic properties.

A threshold must be chosen to distinguish the signal due to fluorescence molecules from the signal due to noise in the background signal. In the data of Fig. 2(A), the threshold was drawn at a signal of 0.1 V amplitude. Those peaks that exceeded

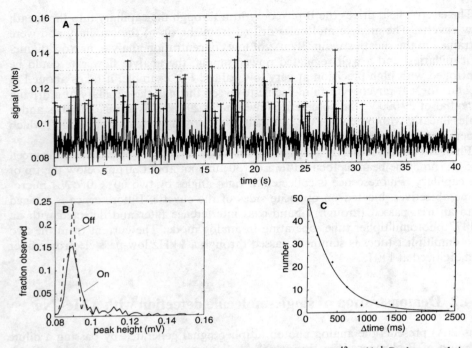

Figure 2. Single-molecule data. A solution containing $1.672 \times 10^{-12}\,\mathrm{mol\,l^{-1}}$ B-phycoerythrin was introduced into a sheath flow cuvette at a potential of 1 000 V, which introduces roughly 6 molecules per second into the system. After 30 seconds, the potential is reversed, generating a background signal for the last 10 seconds of the run. A: Raw fluorescence signal. Those peaks that exceed a 0.1 V threshold are marked with a cross. B: Histogram of peak amplitudes. The dashed curve was generated with the sample off and the solid curve was generated with analyte flowing. C: Time interval histogram. The time between single molecule events was measured and used to construct the histogram. The smooth curve is the least squares fit of an exponential decay to the data.

the threshold are marked with a cross; roughly 3 molecules/second are detected in the experiment. This detection rate is half of the expected rate, and may reflect adsorption of analyte molecules onto the walls of the sample container.

The arrival time of the molecules is random. For a Poisson-distributed population, the time interval between the arrival of successive molecules should follow an exponential decay. Fig. 2(C) presents the probability distribution for the time interval between the arrival of successive molecules, along with the least-squares fit of the data to an exponential decay. In this particular case, the average time between the arrival of molecules is about 350 ms.

Fig. 3 presents similar data, this time generated at 5 kV potential. The number of peaks generated during the analyte flow period is much higher than at 1 kV potential. However, the amplitude of the peaks is similar to that observed in Fig. 2, as expected for single molecule events. About 15 molecules/s are detected as they pass

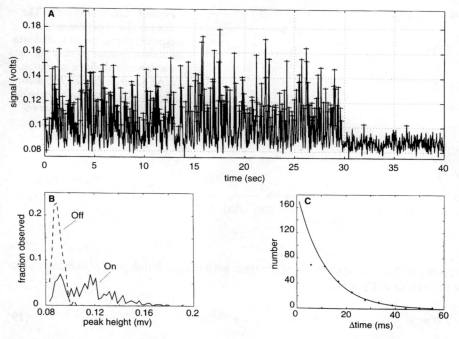

Figure 3. Single-molecule data. A solution containing $1.67 \times 10^{-12}\,\mathrm{mol\,l^{-1}}$ B-phycoerythrin was introduced into a sheath flow cuvette at a potential of 5 000 V, which introduces roughly 30 molecules per second into the system. After 30 seconds, the potential is reversed, generating a background signal for the last 10 seconds of the run. A: Raw fluorescence signal. Those peaks that exceed a 0.1 V threshold are marked with a cross. B: Histogram of peak amplitudes. The dashed curve was generated with the sample off and the solid curve was generated with analyte flowing. C: Time interval histogram. The time between single molecule events was measured and used to construct the histogram. The smooth curve is the least-squares fit of an exponential decay to the data. However, the first time delay datum was not included in the fit; instrument dead-time effects perturb this early time point.

through the laser beam. As in Fig. 2, the detection efficiency is 50%. The histogram shown in Fig. 4(B) is now dominated by the signal from individual molecules, which is well resolved from the background signal. Finally, the arrival-time histogram is shown in Fig. 3(C), along with a least-squares fit to an exponential decay. Relatively few counts are observed for short periods, which reflects the 10 millisecond deadtime of our instrument, during which period we are unable to detect a second molecule.

Fig. 4 presents the calibration curve for the instrument, where the observed number of counts is plotted against the expected number of counts. The curve saturates at 375 counts. The detector is unable to discriminate between one and more than one molecule in the probe volume. The finite dead-volume of the instrument limits the maximum number of counts that can be observed. The smooth curve through the

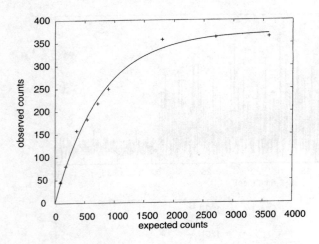

Figure 4. Calibration curve. The number of counts generated and expected during the 30 second measurement time are plotted. The data were compared with a saturation curve based on Poisson distribution. The curve saturates because the detector is unable to resolve fluorescence from more than one molecule present in the probe volume.

data points is the least squares fit of the data with the probability of observing one or more molecules in the detector

$$P(N \geq 1) = 1 - P(N = 0) = 1 - e^{-N/\gamma} \tag{19}$$

where γ the number of molecules per second passing through the detector to cause the saturation. In this case, the detector saturates at a nominal flow rate of 25 molecules/s. Note that the detector responds to only half of the analyte molecules, so that the detector saturates at ca. 12 counts/second. This saturation rate can be increased by reducing the probe volume. However, decreasing the probe volume results in the use of a smaller sample stream, and a concomitant decrease in sample throughput.

3.4.3 Capillary electrophoresis

Finally, my research group has focused on high-efficiency detection for capillary electrophoresis. In our first experiment, a high-power laser was used to excite fluorescence from fluorescein thiocarbamyl amino acids separated by capillary electrophoresis; detection limits of 6,000 molecules were reported [35]. This experiment suffered from photobleaching. The use of a lower power laser reduced detection limits to 600 fluorescein molecules [36]. Similar detection limits were produced for fluorescein-labeled DNA fragments [37]. Work with tetramethylrhodamine-labeled amino acids and a 0.75 mW helium–neon laser produced similar detection limits [45]. The use of cleaner buffers and carefully synthesized samples reduced detection limits to 100 molecules of tetramethylrhodamine labeled oligosaccharides [43]. Appropriate digital filtering of data has reduced detection limit to 30 molecules for rhodamine 6G [46]. The use of a 7 mW helium–neon laser and improved collection optics have resulted in detection limits of 6 molecules of sulforhodamine 101 [47].

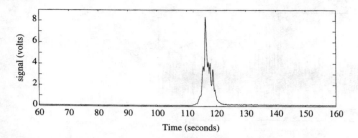

Figure 5. Capillary electrophoresis of 30 000 B-phycoerythrin molecules. A $1.67 \times 10^{-9}\,\text{mol}\,\text{l}^{-1}$ solution of B-phycoerythrin was injected at 1 kV for 1 second onto the 10 µm inner diameter, 29 cm long capillary. The sample was replaced with clean running buffer, and a potential of 29 000 V was applied to the capillary, driving the protein through the capillary to the detector. The structure on the peak is presumably due to different isoforms of the protein; this structure is reproducible from run to run.

Finally, we have reported single molecule detection for B-phycoerythrin, excited with a 2-mW helium–neon laser [40].

A typical electropherogram generated by a relatively large amount of B-phycoerythrin is shown in Fig. 5. There is a relatively broad peak about 120 seconds following the injection. The peak consists of an intense center lobe, surrounded by four or five smaller, partially resolved peaks. At this concentration, the structure of the electropherogram is quite consistent from run to run and presumably reflects differences in the structure of different isoforms of the protein.

Fig. 6 presents an electropherogram generated by the injection of ca. 30 molecules of B-phycoerythrin. The data consist of the noisy background with several sharp spikes. A threshold was set at 0.1 V, and those peaks that exceeded the threshold were identified as single protein molecules and marked with a cross. Fig. 6(B) presents the single-molecule data for this run. The main concentration of molecules falls in the 110–120 second range and was caused by migration of the B-phycoerythrin. There were also three molecules that migrated before the main peak, which represent false alarms caused by either impurity molecules or noise spikes.

In a thought experiment, the same molecule can be injected many times into the capillary and subjected to electrophoretic analysis. The migration time for the molecule will form a probability distribution function, with the width of the distribution associated with diffusion of the molecule in the capillary. In our experiment, diffusion will spread a slug of B-phycoerythrin over a ca. 50 ms time period. The data of Fig. 6(B) represent one sampling of the migration of these molecules from the capillary. We can reconstruct the electropherogram that is expected if a larger population of these molecules was injected onto the capillary by smoothing the data of Fig. 5(B) with a Gaussian function with 50 ms width, Fig. 6(C). Several closely migrating molecules have merged and formed slightly larger peaks, but most of the molecules do not overlap. We can conclude that there are at least eight different types of B-phycoerythrin migrating from this capillary based on the distinct populations that are seen in the single molecule data. This information is not provided in the high

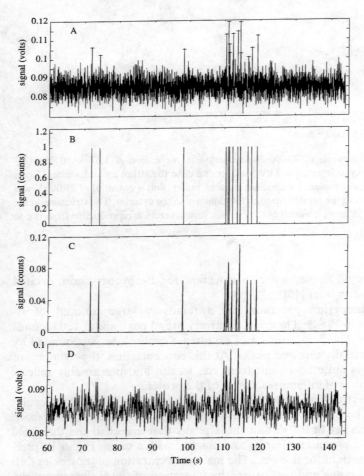

Figure 6. Capillary electrophoresis of about 15 B-phycoerythrin molecules. A $1.67 \times 10^{-12}\,\mathrm{mol\,l^{-1}}$ solution of B-phycoerythrin was injected at 500 V for 5 seconds onto the 10 μm inner diameter, 29 cm long capillary. The sample was replaced with clean running buffer, and a potential of 29 000 V was applied to the capillary, driving the protein through the capillary tot he detector. A: The raw fluorescence signal. Those peaks that exceed 0.1 V are marked with a cross. B: The digitized data, where only those points corresponding to a single molecule are plotted. C: The digitized data after convolution with a Gaussian function with 50 ms standard deviation; this plot simulates the effects of diffusion on the migrating molecules. D: The raw data after passage through the 50 ms filter; most information on the peak is lost.

concentration electropherogram, because the original electropherogram probably contains dozens of fractions that overlap to generate the overlapped set of peaks shown in Fig. 5. Finally, we can filter our original analog signal with a 50 ms filter, and compare the electropherogram with either the high concentration data or the single molecule data; little information can be gained from this noisy signal.

Figure 7. A second injection of ca. 15 molecules. The fluctuation in the peak shape reflects molecular shot noise, which is an irreducible noise source when analyzing small numbers of molecules.

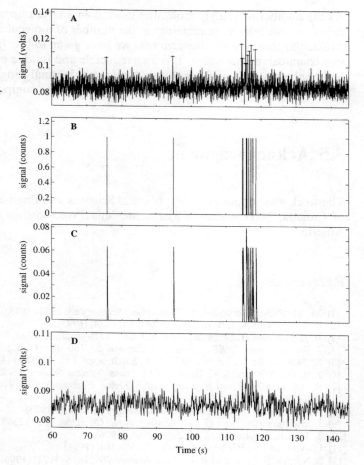

It is important to understand the difference in peak shape between Figs. 6 and 5. In Fig. 5, thirty thousand molecules were taken for analysis, which form a representative sample of the universe of B-phycoerythrin molecules. In Fig. 6, only a small subpopulation of that universe was taken for analysis. This subpopulation is not large enough to form a representative sample of all B-phycoerythrin molecules, and the peak shape is quite different from the parent population. The peak shape changes from injection to injection, as different components of B-phycoerythrin are taken for analysis. Fig. 7 presents another injection, where perhaps seven components of B-phycoerythrin are sampled.

This variation in peak shape between different injections represents a fundamental limitation of chemical analysis at the single molecule level. The distribution in the number of plolecules sampled in any experiment is ultimately limited by Poisson statistics, where the standard deviation of the number of molecules found in any subpopulation is equal to the square root of the average number of molecules taken.

As the number of analyte molecules used in an analysis approaches the single molecule level, the relative uncertainty in the number of molecules present becomes huge. Molecular shot noise is the term that we have given to this fundamental uncertainty in a chemical measurement. It is an irreducible and fundamental limitation of chemical analysis at the single molecule limit; only by analyzing a large sample can we obtain precise information that is characteristic of the composition of the sample.

3.5 Acknowledgment

The work was supported by the Natural Sciences and Engineering Research Council of Canada. NJD acknowledges a McCalla Professorship from the University of Alberta.

References

[1] M. M. Collingson and R. M. Wightman, *Science* **268**, 1883–1885, (1995).

[2] F. R. Fan and A. J. Bard, *Science* **267**, 871–874, (1995).

[3] B. Reiffen and H. Sherman, *Proc. IEEE* **51**, 1316–1320, (1963).

[4] J. W. Goodman, *IEEE J. Quantum. Electron.* **53**, 180–181, (1965).

[5] K. Matsuo, M. C. Teich, and B. E. A. Saleh, *Appl. Opt.* **22**, 1898–1909, (1983).

[6] J. H. Gilchrist and S. K. Babey, *IEEE Trans. Nuclear Science* **NS-25**, 1655–1660, (1978).

[7] R. M. Gagliardi and S. Karp, *IEEE Trans. Communication Technology* **com-17**, 208–216, (1969).

[8] N. J. Dovichi, *Trends in Analytical Chemistry* **3**, 55–57, (1983).

[9] J. M. Hungerford and G. D. Christian, *Anal. Chem.* **58**, 2567–2568, (1986).

[10] T. Hirschfeld, *Appl. Opt.* **15**, 2965–2966, (1976).

[11] U. Mets and R. Rigler, *J. Fluoresc.* **4**, 259–264, (1994).

[12] S. Nie, D. T. Chiu, and R. N. Zare, *Science* **266**, 1018–1021, (1994).

[13] W. B. Whitten, J. M. Ramsey, S. Arnold, and B. V. Bronk, *Anal. Chem.* **63**, 1027–1031, (1991).

[14] K. C. Ng, W. B. Whitten, S. Arnold, and J. M. Ramsey, *Anal. Chem.* **64**, 2914–2929, (1992).

[15] M. D. Barnes, K. C. Ng, W. B. Whitten, and J. M. Ramsey, *Anal. Chem.* **65**, 2360–2365, (1993).

[16] M. D. Barnes, W. B. Whitten, and J. M. Ramsey, *Anal. Chem.* **67**, 418A–423A, (1995).

[17] N. J. Dovichi, J. C. Martin, J. H. Jett, and R. A. Keller, *Science* **219**, 845–847, (1983).

[18] N. J. Dovichi, J. C. Martin, J. H. Jett, M. Trkula, and R. A. Keller, *Anal. Chem.* **56**, 348–354, (1984).

[19] D. C. Nguyen, R. A. Keller, J. H. Jett, and J. C. Martin, *Anal. Chem.* **59**, 2158–2160, (1987).

[20] J. H. Hahn, S. A. Soper, H. L. Nutter, J. C. Martin, J. H. Jett, and R. A. Keller, *Applied Spectroscopy*, **45**, 743–745, (1991).

[21] P. M. Goodwin, M. E. Johnson, J. C. Martin, W. P. Ambrose, J. H. Jett, and R. A. Keller, *Nucleic Acid Research* **21**, 803–810, (1993).

[22] A. Castro, F. R. Fairfield, and E. B. Shera, *Anal. Chem.* **65**, 849–852, (1993).

[23] B. B. Haab, and R. A. Mathies, *Anal. Chem.* **67**, 3253–3260, (1965).

[24] E. B. Shera, N. K. Seitzinger, L. M. Davis, R. A. Keller, and S. A. Soper, *Chem. Phys. Letters* **174**, 553–557, (1990).

[25] S. A. Soper, E. B. Shera, J. C. Martin, J. H. Jett, J. H. Hahn, H. L. Nutter, and R. A. Keller, *Anal. Chem.* **63**, 432–437, (1991).

[26] C. W. Wilkerson, P. M. Goodwin, W. P. Ambrose, J. C. Martin, and R. A. Keller, *Appl. Phys. Lett.* **62**, 2030–2032, (1993).

[27] K. Peck, L. Stryer, A. N. Glazer, and R. A. Mathies, *Proc. Natl. Acad. Sci. USA* **86**, 4087–4091, (1989).

[28] S. A. Soper, Q. L. Mattingly, and P. Vegunta, *Anal. Chem.* **65**, 740–747, (1993).

[29] S. A. Soper, L. M. Davis, and E. B. Shera, *J. Opt. Soc. Amer.* **65**, 1761–1769, (1993).

[30] Y. H. Lee, R. G. Maus, B. W. Smith, and J. D. Wineforner, *Anal. Chem.* **66**, 4142–4149, (1994).

[31] J. H. Jett, R. A. Keller, J. C. Martin, B. L. Marrone, R. K. Moyzis, R. L. Ratliff, N. K. Seitzinger, E. B. Shera, and C. C. Stewart, *J. Biomol. Struct. Dynamics* **7**, 301–310, (1989).

[32] L. M. Davis, F. R. Fairfield, C. A. Harger, J. H. Jett, R. A. Keller, J. H. Hahn, L. A. Krakowski, B. L. Marrone, J. C. Martin, H. L. Nutter, R. L. Ratliff, E. B. Shera, D. J. Simpson, and S. A. Soper, *Genetic Anal.* **8**, 1–7, (1991).

[33] J. W. Jorgenson and K. D. Lukacs, *Science* **222**, 266–272, (1983).

[34] J. Landers (Ed.) *CRC Handbook of Capillary Electrophoresis*, CRC Press, Boca Raton, 1993.

[35] Y. F. Cheng and N. J. Dovichi, *Science* **242**, 562–564, (1988).

[36] S. Wu and N. J. Dovichi, *J. Chromatogr.* **480**, 141–155, (1989).

[37] H. Swerdlow, J. Z. Zhang, D. Y. Chen, H. R. Harke, R. Grey, S. Wu, N. J. Dovichi, and C. Fuller, *Anal. Chem.* **63**, 2835–2841, (1991).

[38] J. Z. Zhang, Y. Fang, J. Y. Hou, H. J. Ren, R. Jiang, P. Roos, and N. J. Dovichi, *Anal. Chem.* **67**, 4589–4593, (1995).

[39] D. M. Pinto, E. A. Arriaga, S. Sia, Z. Li, and N. J. Dovichi, *Electrophoresis* **16**, 534–540, (1995).

[40] D. Y. Chen and N. J. Dovichi, *Anal. Chem.* **68**, 690–696, (1996).

[41] D. Craig, E. Arriaga, P. Banks, Y. Zhang, A. Renborg, M. M. Palcic, and N. J. Dovichi, *Anal. Biochem.* **226**, 147–153, (1995).

[42] D. B. Craig, J. C. Y. Wong, and N. J. Dovichi, *Anal. Chem.* **68**, 697–700, (1996).

[43] J. Y. Zhao, N. J. Dovichi, O. Hindsgaul, S. Gosselin, and M. M. Palcic, *Glycobiology* **4**, 239–242, (1994).

[44] Y. Zhang, X. C. Li, N. J. Dovichi, C. A. Compston, M. M. Palcic, P. Diedrich, and O. Hindsgaul, *Anal. Biochem.* **227**, 368–376, (1995).

[45] J. Y. Zhao, D. Y. Chen, and N. J. Dovichi, *J. Chromatogr.* **608**, 117–120, (1992).

[46] D. Y. Chen and N. J. Dovichi, *J. Chromatogr.* **657**, 265–269, (1994).

[47] D. Y. Chen, K. Adelhelm, X. L. Cheng, and N. J. Dovichi, *Analyst* **119**, 349–352, (1994).

Index